EASY PLANT PROPAGATION

A Simple Homemade Plant Propagation System

By

Michael J. McGroarty

for McGroarty Enterprises Inc.

Bloomington, IN Milton Keynes, UK

AuthorHouse™
1663 Liberty Drive, Suite 200
Bloomington, IN 47403
www.authorhouse.com
Phone: 1-800-839-8640

AuthorHouse™ UK Ltd.
500 Avebury Boulevard
Central Milton Keynes, MK9 2BE
www.authorhouse.co.uk
Phone: 08001974150

First published by AuthorHouse 1/15/2007

ISBN: 978-1-4259-8570-7 (e)
ISBN: 978-1-4259-8568-4 (sc)
ISBN: 978-1-4259-8569-1 (hc)

Library of Congress Control Number: 2006910896

Printed in the United States of America
Bloomington, Indiana

This book is printed on acid-free paper.

This book is dedicated to my wife Pam.

The year was 1970. The season, Halloween. Shy and quiet, wandering around in the old barn, sipping cider, trying to look cool like the rest of my fifteen year old classmates at the party. Too shy to actually speak to any of the girls, when suddenly Wally grabs me by the arm, walks me over to a bale of straw and forces me to sit on the bale next to a very pretty young lady who was even more shy than I was.

It was a set up.

She was sitting off to the side on the end of the bale, just waiting for Wally to bring the victim by. Nobody sits off to the side on the end of a bale of straw.

I should have seen it coming!

Over the next 36 years, the bond that started on a bale of straw in an old cow barn would take us on a journey that we couldn't have predicted in a million years. As I struggled to make my way in the world with one brainstorm after another, she was there by my side. We've been flat broke, we've prospered, we've seen some successes, and even more failures.

We raised two incredibly bright sons and when I look at them I see their mother. Not just in appearance, but in their character, integrity, independence, and a million other quality traits she methodically instilled in them.

When I look at her I see the love of my life!

Contents

A message from the author . . .

I'd like to thank you for purchasing this book. A lot has happened since I published the first version of this book back in 1996. At that time I wasn't even sure if I could sell enough books to break even, let alone make a profit.

Then in 1999 I created my http://freeplants.com/ website and the first version completely sold out and is now out of print.

When my website was fairly new I received an Email from a gentleman in Australia. He wrote, "I really love your website, but when you explain how to do this and that I wish you would mention the season and not the month of the year, because here in Australia our seasons are the exact opposite of yours and it confuses me." I told him that I would keep that in mind, and I hope I do as I rewrite this book. Then I went on to explain that when I wrote my original gardening articles I wasn't even sure if anybody down the road would read them, let alone people on the other side of the world.

It has been an adventure!

Growing your own landscape plants from scratch is much easier than you might think. Many of the techniques that you will learn in this book are so simple that young children can do them, and with great success! Finding a beautiful landscape plant and being able to reproduce that plant as many times as your heart desires is not only fun and exciting, but can result in a tremendous amount of savings and/or earnings.

I first became acquainted with landscape plants as a high school student working afternoons, evenings, and weekends at one of the larger wholesale nurseries in the community, often putting in 40-

50 hours per week while still in high school. At that time I had no interest in landscape plants. It was just a job, but I was delighted to have it. Who wouldn't have been delighted? I was earning $1.60 per hour!

As I laboriously planted, weeded, watered, potted, carried, pruned, loaded, counted, packaged, tied, and dug landscape plants, I slowly learned their common names, botanical names, and many of their unique characteristics. I still did not have much interest in landscape plants, but I was becoming somewhat of an unwilling expert.

After high school graduation I went to work for a landscape contractor whom I had met at the nursery. I wasn't any more interested in landscaping than I had been in plants, but he had construction equipment, front-end loaders, bulldozers and such, and I wanted to learn how to operate heavy equipment.

After a few weeks something strange began to happen. I actually started to enjoy landscaping. It was most satisfying to see the plants that I had been working with for so long being artistically placed around peoples' homes. I also realized that I knew more about landscape plants than my boss did. I figured if he could landscape homes with his limited knowledge of plants, then so could I.

With that, a career was born, and in the thirty-some years since, I have never ventured far away from plants, horticulture and the nursery business. Today on the internet I teach people from around the world how to take a very small area in their backyard and turn that into a viable backyard enterprise. I am very proud to say that my efforts have literally changed peoples' lives.

Propagating plants is addictive, and at http://freeplants.com/ we have a worldwide community of people who share this passion and exchange ideas daily on our message board. If you have a question, stop by and you'll get an answer on the board. At the time of this writing, our message board is available to the public, and I do hope that we can keep this board open for a long time. Message boards are an incredible amount of work, but I will do my best to keep it open for you, my gardening friend. Please do pay us a visit. http://freeplants.com

From those early days when I realized that I enjoyed landscaping, I have slowly, but surely, developed a lifetime love affair with landscape plants. Landscape plants have been part of my daily life for the past thirty-three years, they have put food on the table and tennis shoes on the kids. Working in the landscape/nursery industry has been incredibly rewarding for me.

It is my sincere hope that through the pages of this book, you too will learn and develop a hobby that will bring you just as many rewards, and just as much enjoyment. Or do as many of my readers have done, take this hobby to the next level and grow and sell small plants at a very nice profit.

Sincerely,
Michael J. McGroarty
http://freeplants.com

Notice: Plant Patents and Trademarks

Some plants are patented and it is against the law to reproduce those plants unless you have purchased a license from the holder of the patent. Then you must pay royalties to the patent holder. Licenses usually start at $1,000 or more, and require that you keep detailed records of what you produce and sell.

Other plants are protected with a registered trademark. The same rules apply as above.

In today's plant marketplace we are seeing more and more patented and registered plants being marketed and sold. Why is that? In my opinion, and this is just my opinion, it is because it makes good business sense for those selling plants on the wholesale level. Why would a grower heavily promote plants that others are free to propagate, when instead he can heavily promote and sell plants that produce a perpetual income through royalty payments? From a business sense it is perfectly sound reasoning.

But for you and I who like to tinker around at home propagating plants, or for those who have purchased my **Backyard Growing System from http://freeplants.com,** it is extremely important for us to remember that there are hundreds, if not thousands of plant varieties that we are free to propagate. So don't get all caught up trying to find a way to propagate a protected plant without getting in trouble. Forget about those protected plants and reproduce the beautiful plants that you are free to propagate. In large part, the beautiful plants that served us well for so many years are being kicked aside by the industry to make room for royalty-producing plants.

This is something that I constantly remind my **Backyard Growers**. Many folks have taken my **Backyard Growing System** and put it into action, they've built the intermittent mist plant propagating system that I show them how to build and use, and they are propagating plants by the thousands. It's imperative that they find a steady supply of new plants that they are free to propagate. By new plants I really mean old plants that are being kicked to the curb by the industry. They really aren't old plants at all. Sure, they've been around for generations, but the generation of today has never seen them before. They will happily enjoy them just as much as any other plant on the market.

In an effort to combat the confusion of "What can I propagate and what can't I propagate?", my **Backyard Growers Group** now sells rooted cuttings and other small plants to each other to be used as stock plants. They make a concentrated effort to find really nice plants that are not patented or trademarked, they start reproducing the plants they find, and offer them for sale to each other. Of course they also grow them on to larger sizes and sell them to retail and wholesale customers. It's a great way to acquire beautiful plants that they are free to propagate. It's also a great way to buy plants in general because they sell them to other nurseries and **Backyard Growers** for as little as 67¢ each.

In short, forget about patented plants and plants with trademarked names. Believe me, there is an almost endless supply of beautiful plants that you are free to propagate.

How to Use this Book

This book is divided into chapters, all with unique information. First you want to read the entire book, and then reread it as often as you can because each time you read the entire book you will understand the process of plant propagation better, picking up on things that didn't make perfect sense to you the first time.

There are chapters that you can reference each time you try to propagate a new plant, or each time you attempt plant propagation in a different season of the year. Yes, you can propagate plants year-round; it's only a matter of applying the correct propagation technique at the correct time of the year. Near the back of the book you will find a chapter titled, "What You Should Be Doing Now", which gives you a month by month schedule of what plant propagation techniques work the best at that time of the year. Make a mental note of this chapter, or better yet, mark your calendar with reminders so you can experience the joy of gardening all year.

Near the back of the book you will also find a chapter titled, "How to Do What". In that chapter you will find a list of some of the most popular landscape plants, with a description of what propagation techniques work the best for that particular plant and the time of the year to employ that propagation technique. This particular chapter does not include information for every plant imaginable, but you can always go to the message board at http://freeplants.com and ask about a particular plant.

Near the front of the book you will find the chapter, "The Basics of Plant Propagation". This chapter will help you understand the difference between softwood cuttings and hardwood cuttings and when you should use each. It will also help you determine what category the plant you wish to propagate falls in. Is it deciduous, evergreen, broadleaf evergreen, multi-crown, etc?

Then you will find "How to Build a Simple, Homemade Plant Propagation System". This chapter gives you step-by-step instructions and photos on how to build and use this simple system. The homemade plant propagation system detailed in this book makes plant propagation incredibly easy. It's not something that you necessarily use for every technique in the book, but it does enhance the success you'll see using many of the strategies in the book.

You will also find a chapter on how to care for your baby plants once you have them rooted. This chapter covers some of the basics of fertilization and how to make sure your little plants make it through their first winter.

There are also separate chapters for each plant propagation technique detailed in the book so you can quickly review any technique you need to without thumbing through the entire book.

There are not a lot of photos of the different plant propagation techniques in the book, but you will find plenty of color photos on my website, and I am currently in the process of adding some short video clips of various plant propagation techniques to my website.

There you have it. This book will give you an incredible foundation as a plant propagator, and any additional information, including the ability to ask questions, can be found on my website, http://freeplants.com

Basic Information about Landscape Plants That Will Make You a More Successful, and Happier Gardener

In a period of about 30 years, I've spent a great deal of time working with and consulting people about gardening issues and I've seen hundreds of people making the same gardening mistakes over and over. Oftentimes simply because they are given bad information from people they should be able to trust. So in this chapter I will lay out the basics of what it takes to make your landscape happy and thrive.

Landscape plants love good rich topsoil.

Landscape plants love well-drained soil.

Landscape plants need water, but they don't particularly love water.

They just need enough moisture on a regular basis to keep them healthy and happy. Think about those towering trees in the forest. Who waters them? Mother Nature. Usually on an as-needed basis. In the event of a drought they struggle, and at times the small ones perish. It is only in those periods of drought do they wish that somebody had intervened. The plants in your yard are pretty much the same. When first planted, they only have a limited root system and need to be watered once or twice a week. Then as they become more established they require less water.

It is extremely important that they be planted in well-drained soil so when they do receive water, any excess water can quickly drain away from the root system. If the soil in your yard is clay soil that drains poorly, the best thing you can do is build raised beds using good rich topsoil. A raised bed is much better than trying to amend your clay soil because when you add organic and other porous matter to clay soil, you can improve the quality of the soil, but you actually make the drainage problem worse.

When you take one section of your yard and till in porous material, you enable water to drain into that area easily, but there is no way for it to drain through, which is what you really need to happen. So in essence, you create a bathtub that retains water and many of your plants will die from a lack of oxygen because they are submerged in water for long periods of time.

Plants need oxygen to their root systems, and they actually draw that oxygen through the soil, not the foliage. This brings me to another important point, a mistake that people make way too often.

Plants will not tolerate being planted too deeply!

People just don't know any better, and they make the mistake of installing trees and shrubs too deep in the ground, and the plants die from being too wet, and/or a lack of oxygen. I see this all the time, and people lose beautiful, very expensive plants because they didn't know any better, or they were given poor advice by someone who shouldn't be administering gardening advice.

When you purchase a plant, be it in a container or balled in burlap, the plant already has an established crown. The established crown is either the top of the soil in the container, or the top of the root ball. That crown should be planted at or above grade. In most cases I plant with the crown slightly above grade, then mound an inch or two of soil over the crown, then mulch over that. This allows the plant to adequately breathe.

Plants don't like fertilizer.

As a matter of fact, they hate it. Yes, there are times when you need to fertilize plants. Plants growing in containers must be fertilized because they need to be watered daily, and this constant watering leaches all of the nutrients out of the growing medium. But you cannot use regular garden fertilizer on plants growing in containers. You can feed them with a liquid fertilizer that you mix with water and spray on the foliage, or you can sprinkle a granular slow-release fertilizer on the surface of the soil in the container. But make sure the granular fertilizer you use takes three to four months to completely release.

But please understand . . . plants prefer to be planted in the ground, not grown in containers. In the ground they are easier to care for, and they will thrive. I'll write more about that in another chapter.

Around your house, the plants in your landscaping should need very little, if any fertilizer at all. If you have a plant that is really doing poorly, chances are it is too wet, too dry, or is being affected by some kind of pest. More than likely it does not need fertilizer, and if you fertilize a plant that is showing signs of a struggle to survive, you'll probably do it in! As long as your plants are planted in good rich soil that drains well, they should do just fine.

If you've been to my http://freeplants.com website you have probably seen photos of my backyard. People comment constantly about those photos and how beautiful everything looks. My backyard is a perfect example of what you've just read. The soil on that property is all sand and gravel. There are only a few inches of topsoil, and even that is mostly sand and gravel. When I built the mounds that you see in the photos, I rented a machine and excavated areas where I wanted my nursery, and I just took that excavated soil and pushed it up into piles that we later landscaped. There is very little topsoil in any of those mounds, mostly sand and gravel.

When we landscaped, we simply dug holes and dropped the plants in. No fertilizer, no peat moss, no compost, cow manure, bone meal, or any other soil amendments that they love to sell you at the garden center. We just dug holes and put the plants in. Not once did I ever fertilize those plants. Look at the photos. The plants thrived from day one, planted just as nature intended.

Read this chapter again. When your next-door neighbor, who seems to know everything, tries to tell you differently, let him or her read it. Just don't let him keep your book!

This is the one area that I will stand my ground and claim to be the expert. Over the years I've installed tens of thousands of plants and I've guaranteed those plants to live, and I've backed up that guarantee with my grocery money. In my book, that makes me an expert.

On to the fun stuff!

How to Build a Simple, Homemade Plant Propagation System

The homemade plant propagation system that I'm about to show you how to build is very easy to build, and it's very inexpensive. This is one of the most fun and rewarding things you can do in your garden. You'll be amazed at the number of plants you can propagate using this simple system.

Photo 1. Taping up the aquarium.

The first thing you need to locate is a new or used fish tank, or fish aquarium if you prefer. You can use any size you like, but if you keep it on the smaller size you may find it easier to work with. I prefer to build more than one of the smaller versions, rather than wrestle with a large one. You can often find used fish aquariums at garage sales and flea markets for a dollar or two. The one you see in **Photo 1** I purchased new at one of the discount pet stores for less than ten dollars. Notice how Dana is applying masking tape to the aquarium? I'll explain why in a minute.

Once you have your fish aquarium, you need to measure to get the outside dimensions of the aquarium. Length and width are all you need; don't worry about measuring the height of the aquarium. Keep in mind, during the building of your Propagation System you should wear work gloves and eye protection.

Photo 2. Building the propagation flat.

Next using 1- by 4-inch lumber you are going to construct a box, or a flat with an open bottom, like the one Dana is working on in **Photo 2**. Make the inside of your flat one inch larger than your aquarium in both directions. In other words, if you measure across the top of your aquarium and find that the width is 9½ inches, make the width of your flat 10½ inches on the inside. If the length is 20 inches, make your box 21 inches on the inside.

When your flat is complete you should be able to turn the aquarium upside down and place it inside the flat and have approximately ½-inch of space on all four sides.

Photo 3. Lining the flat with hardware cloth.

Notice how simple the flat is to build? It requires just four pieces of wood to make the four sides, and three pieces across the bottom. Notice that there is a small gap at each end of the bottom and two larger gaps in the middle? The size of these gaps isn't critical as long as they are at least one inch each. They are there to allow water to drain through so your cuttings and rooting medium don't stay too wet.

It's difficult to see in **Photo 3,** but if you look closely you'll see that we lined the inside of the bottom of the flat with a piece of rigid hardware cloth (screen). Make sure the hardware cloth you use is quite rigid so it doesn't sag below the gaps on the bottom when the flat is full of damp potting soil or rooting medium. If you go to a full service hardware store they will have what you need.

Photo 4. Painting the aquarium.

Now it's time to paint the homemade plant propagation system. From now on the aquarium will be used upside down, so what was intended to be the bottom is now going to be the top. Confusing, I know, but the photos should help.

Notice how Dana applied masking tape to the top of the aquarium? You'll see in **Photo 5** that she applied a strip of masking tape, left a one-inch gap, then another strip of masking tape, and another one-inch gap, then another strip of tape and so on. In **Photo 4** you can also see that she applied a piece of masking tape all the way around the aquarium, about one inch from the top. If you look closely you'll see that she left the ends of the tape sticking out past the end of the aquarium. This is so you can quickly and easily remove the tape after spray painting the aquarium.

Once you have the aquarium taped up as shown in **Photo 4,** you will then paint the outside of the aquarium with white paint, painting over the masking tape and any exposed areas of the aquarium. The paint must be white, and make sure the paint you buy will adhere to either glass or plastic, depending on what your aquarium is made out of. We use white paint to reflect the rays from the sun so our plant propagator does not get too hot.

After painting the outside of the aquarium you will need to remove the masking tape, and you want to do so as soon as the paint becomes tacky. If you wait until the paint is dry it will stick to the masking tape and you'll pull up paint that is not supposed to be removed.

Photo 5. After painting the aquarium, the masking tape is removed.

In **Photo 5**, with the aquarium painted and the masking tape removed, you can see that we now have strips of clear plastic exposed. These clear strips allow a minimal amount of sunlight to enter our plant propagation system, while reflecting the majority of the rays from the sun. This creates as close to the ideal environment for plant propagation as we can get at home.

Photo 6. Filling the flat with growing medium.

Now it's time to fill the flat with the growing medium of your choice. You can use sand or bagged potting soil, but you have to be careful not to buy potting soil that has nitrogen and other elements added, which is the case with a lot of potting soils. During the rooting process you do not want to encourage top growth on your cuttings while they are rooting; thus no fertilizers during the rooting process. Another problem with many potting soils is poor drainage, so it's advisable to mix in some perlite or vermiculite to loosen the soil and enhance the drainage. Either will work just fine. You can also use a mixture of peat moss and perlite or vermiculite which seems to work quite well, and peat is neutral (it has insignificant amounts of nitrogen and other elements). Mix one part peat moss to three parts perlite for best results.

When rooting cuttings you want the growing medium moist but not wet and soggy. It's like baking a cake. Nobody likes a dry cake, and most people prefer a moist cake. But nobody wants a wet or soggy cake. Your cuttings prefer their growing medium the same way; moist, but not wet and soggy.

Fill the propagation flat all the way to the top with the growing medium of your choice, and pack the medium down. You can even water it in to make sure all of the air pockets are removed and the flat is completely full. You want the growing medium firm because the cover (the aquarium) rests on the growing medium, not the wooden frame.

Photo 7. Sticking cuttings is a simple process of clip, strip, dip, stick and water.

That's all there is to it. Your Homemade Plant Propagation System is complete and ready to accept the cuttings of your choice. I won't go into great detail about how to actually make the cuttings here because that will be covered elsewhere in the book, but I will give you a few tips on how to use the propagator you built. The types of plants that you can root or germinate using this propagator make up an almost endless list. You don't really know for sure what will root and what won't root until you try.

For the most part, shrubs of all kinds - deciduous, evergreen, and broadleaf evergreen - are propagated from cuttings. Many trees are grown from seed. In the back of the book there is a chapter titled "How to Do What" with a list of many popular plants and how they are typically propagated.

I encourage you to try all kinds of plants, and I hope you'll visit the message board at http://freeplants.com to learn all you can, and share your successes and failures with me and others. At my website you will also find "how to" videos on all kinds of gardening tasks, including making and sticking cuttings at different times of the year.

Preparing a cutting before sticking it is really simple. Just take a four- or five-inch cutting from the tip of a branch, strip the leaves or needles from the bottom half of the cutting, dip the cutting in a rooting compound, make a little hole in the rooting medium with a screwdriver, ice pick, or even a putty knife, slide the cutting into the hole, and after you have one row of cuttings in place, lightly pack the rooting medium around the cuttings to remove any air pockets.

**Photo 8. Place the cover on your propagator
and the rooting process begins.**

Once you have the flat completely full, give the cuttings a good watering to remove any air pockets that might remain and place the cover on the propagator. That's it! Then all you have to do is keep an eye on the propagator, making sure that there is water beaded up on the inside of the cover at all times. If it looks a little dry, remove the cover and water again.

Rooting compounds come in both liquid and powder form. Both work equally well, but I find the liquid much easier to use because you need different strengths of rooting compound at different times of the year. With powders, they sell different strengths and they don't always have the one you need. With liquid, you mix the rooting compound with water, so all you have to do is follow the instructions on the bottle and mix the strength you need. You can buy rooting compounds at most full service garden stores, or it's easy enough to find online. Just type "liquid rooting compound" into your favorite search engine and I'm sure all kinds of sources will appear. Brand name really doesn't make a difference; they all work pretty much the same.

Keep in mind, rooting compound is not the magic or the secret to successfully rooting cuttings. Success has a lot more to do with timing and technique. Doing the cuttings at the correct time of the year, and applying the proper technique for that time of the year is what will bring you the most success. By the time you finish this book you should have a pretty good understanding of how to do what, at what time of the year.

Some cuttings will root in a matter of days; other can take weeks, if not months. You will learn more about the specifics of rooting different types of cuttings in a later chapter.

How to Care for Your Plants
Once You Have them Rooted

Once you have your plants rooted, you can gently remove them from the propagator by simply slipping your fingers into the growing medium and carefully sliding them under the rooted cuttings, making sure the cuttings are loose from the screen in the bottom of the propagator. Before starting to remove the rooted cuttings, you should tip the propagator up on one end, looking underneath to see how many, if any, roots have grown through the screen. If you leave your rooted cuttings in the propagator too long, the roots will eventually find their way through the screen and that will make them more difficult to remove.

If you have a significant amount of roots growing through the screen, it's best to take a knife or pruning shears and clip those roots off flush with the screen before trying to pull them up through the screen from the top. However, clipping off some small roots really won't be a problem, but if you have to clip off a lot of roots you could shock the plants, unless of course they happen to be dormant at the time. If you have cuttings that have been in the propagator for a long time and are really well rooted, it might be advantageous to wait until the rooted cuttings are dormant before trying to remove them.

Dormancy in the fall occurs after the plants have been subjected to a hard freeze where the temperature dips below 32 degrees Fahrenheit for a period of several hours. After experiencing this type of hard freeze, plants go dormant and will remain dormant until spring. You can safely transplant your cuttings in the fall after they are dormant, or leave them in the flat until early spring and transplant them before they come out of dormancy.

You can safely transplant cuttings during the growing season as long as you don't cut or break a lot of roots in the process. If you remove your cuttings from the propagator shortly after they have rooted, you'll be fine. It's only when the roots have become entangled or are growing through the screen on the bottom of the propagator that it becomes necessary to wait for dormancy.

That's why it's important to remove the cuttings as soon as they are well rooted. Checking the bottom of the screen after you know your cuttings have started to root is a good idea, and then you'll know when it's time to remove them. After your cuttings have been in the propagator you can check rooting progress by gently pulling on the cuttings. The growing medium you are using should be light enough that a cutting without roots can be easily pulled from the medium without damaging any small roots that have started to form. If you feel any resistance at all, stop pulling. Your cuttings are starting to root.

After your cuttings are well rooted, the place to put them is in a bed of good rich soil. If you have really poor soil at your house, you should build a raised bed with good rich topsoil and/or composted material. The bed should be raised at least eight inches. I don't recommend putting your rooted cuttings in pots for a number of reasons. Growing plants in containers is a lot more difficult than growing plants in a bed of good rich soil. Container-grown plants have to be watered every day, sometimes twice a day. The potting soil you use has to be good quality, and it has to be well drained. The daily watering of container-grown plants quickly leaches the nutrients from the potting soil and those nutrients must be replaced through fertilization. Applying fertilizers to container- grown plants is tricky, and it's easy to do more harm than good.

That's why I recommend growing your cuttings in a bed for a year or two after rooting. You can almost plant them and forget about them. Well, after a few weeks you can forget about them.

Michael J McGroarty

Planting your rooted cuttings in a bed is a very simple process. The soil should be loosened before you start planting. Then it's as simple as digging a small hole with a trowel, placing the rooted cutting in the hole, backfilling around the plant with soil, and lightly packing the soil around the roots to remove any air pockets. You don't need peat moss, bone meal or any of the other concoctions they try to sell you at the garden center. **Least of all, you don't need fertilizer!** If you plant your rooted cuttings in good rich soil, they don't need fertilizer. They will grow just fine without it, exactly as Mother Nature intended.

The rule of thumb about fertilizing is this: Not enough is always better than too much.

Planting depth is important, so read the next paragraph carefully. This is the most important paragraph in this book.

All plants - trees, shrubs, and rooted cuttings - have a root crown. The root crown is the place on the plant that separates the top of the plant from the root zone. If you hold up a bare root tree, shrub or rooted cutting, you can easily see where the roots start. That is the root crown. Everything above that point should be above ground, and everything below that point should be below ground.

If there is one gardening mistake that occurs more often than all of the other gardening mistakes combined, it would be planting trees and shrubs too deep in the hole. A lot of people are completely unaware that plants have a root crown, nor do they understand that getting a plant installed in a planting hole at the correct depth is critical. They just dig a hole and drop the plant in the hole, and all too often the plants are too deep and they either drown or suffocate from a lack of oxygen.

Plants need a constant supply of oxygen to their roots, and they get that oxygen from the air. The roots must be close enough to the surface of the soil for the transfer of oxygen to take place.

So it is with rooted cuttings. They too must be planted at the correct depth, and since they are such tiny plants, the correct depth is not very deep at all.

Keeping your rooted cuttings watered for the first few weeks is really important. You don't want to make the soil soggy, but you must keep in mind that the entire root zone is just an inch or two below the surface of the soil, and the sun dries out the top inch or two of the soil surface quickly. So it is important to water your rooted cuttings as needed, at least once or twice a day for the first one or two weeks after planting. After the rooted cuttings have been planted for several weeks, they will need less and less water, to the point that they will be almost completely self sufficient except in the case of drought.

Keep Your Rooted Cuttings Trimmed

Regular and proper pruning of your rooted cuttings is essential for them to mature into full and healthy landscape plants. The more you trim a plant the fuller it gets, and it's important to start this pruning process soon after you plant your rooted cuttings. Soon after you plant your rooted cuttings in a bed they will start to put on new growth. As exciting as it is to see the plants you grew with your own two hands starting to grow, it's important to control that growth so your rooted cuttings develop into beautiful plants.

Plants have two types of buds. Terminal buds are the buds that appear at the tip of the branch, and lateral buds are the buds that appear along the stem, lower than the terminal bud. Terminal buds tend to grow straight out, not stopping until they are either clipped manually, or they are stopped by the end of the growing season. Of course all plants grow at different rates, and some plants only grow from two to four inches in a season, where others can grow as much as three feet. It is these faster growing plants that really need pruning on a regular basis.

Of course, trimming to produce a tree is very different than trimming to produce a shrub, so let's discuss shrub pruning first. Understand that all plants need sunlight to grow properly, and most really enjoy sunlight. The more sun they get the better they grow. But as plants grow they also tend to shade themselves. Most shrubs grow in a vase shape with the top wider than the bottom. This pattern of vase-shaped growing means that the top of the plant shades a great deal of the bottom of the plant. That's why it is so important to start pruning your plants early, so you can force them to be nice and full at the bottom, before the top of the plant has an opportunity to shade the bottom of the plant.

As your rooted cuttings start to grow, the terminal buds have a plan of their own. Their plan is to reach for the stars and to get there as quickly as possible. You have to interrupt this plan by allowing the terminal bud to start growing, but as soon as that new growth is four to six inches long, clip the tip of that branch off. What this does is stop that "I'm reaching for the stars" plan, and it forces the plant to stop growing momentarily and regroup. During this regrouping process what the plant does is replace the terminal bud that you removed, not with one terminal bud, but with multiple terminal buds that all have the same plan as the terminal bud that you removed.

Your job is to once again interrupt the plan by allowing these new terminal buds to each put on four to six inches of new growth, and then once again clip off the tip of each of these new branches. The plant once again regroups and forms even more terminal buds, each destined to reach the stars. Through this constant pruning your plant does grow, but instead of growing tall and lanky, you force it to become nice and full at the bottom, creating a very nice landscape plant. The diagram below illustrates this.

Diagram 1. The plant on the left represents a shrub that has not been pruned at all. Notice how it is growing tall, but very thin inside. The drawing in the center shows how the plant should be pruned, and the drawing on the right shows what the plant should look like after proper pruning.

Trees grown from cuttings or seeds also need proper pruning. But in the case of a tree that is to be grown as a single-stem tree, as most but not all trees are grown, you have to make certain that the tree only has one stem (branch) growing upright. When the tree is young and small it's okay to leave some of the lower branches on the tree. The leaves on these lower, lateral branches help to feed the tree as it grows. But as these branches start to get larger, say 3/16-inch in diameter, they should be removed, cutting them back to the main stem. Make sure you make the cut flush with the stem and don't leave a stub sticking out.

Allow the single stem branch with the terminal bud to grow to the height where you want the tree to start branching. With many ornamental trees this would be about 42 inches from the ground. When the single-stem terminal branch of your tree reaches the desired height, just clip the top off. This will force the tree to stop growing momentarily, and it will more than likely set multiple buds just below where you made your cut. These multiple buds will form the head of the tree, and you need to prune them much the same as you would a shrub, forcing the tree to grow tight and full. Even if the tree only sets one bud below where you made your cut, you need to start the pruning process so your tree grows full.

Once you remove the terminal bud at the desired height and new branches begin to form, it is then time to remove all of the lateral branches that are growing from the stem of your tree so you have a single stem tree with a head that is starting to form.

If your tree cutting or seedling has two stems, each growing from the base at an angle, you must select the strongest and straightest of these two stems to serve as the main stem of your tree, and remove the other one completely. It is often also necessary to stake young trees to force the stems to grow straight, perpendicular to the ground. Do this when the tree is very young. Don't wait thinking you can always do it later. You need to stake young trees as early as possible.

You can buy stakes at a garden center, or you can use half-inch electromagnet tubing (conduit). Just tape the tree to the stake using several wraps of masking tape. The masking tape won't last long in the weather, but that's part of the reason I recommend it. You can use duct tape or electrical tape, but it doesn't fall off on its own and if you forget to remove it, the tree will grow over top of it, seriously damaging the stem of the tree.

Transplanting from the Growing Bed

Once you have your rooted cuttings or seedlings successfully growing in a bed of rich soil, they will quickly become rooted in and start growing like crazy. At some point you will move them from the growing bed to be either planted in a permanent location, or possibly potted so they can be sold or safely transported to a different location. Keep in mind that you should only move plants from your growing bed when they are dormant. Moving them during the growing season can easily shock them, and more than likely kill them. All transplanting should be done when plants are dormant.

As mentioned earlier, dormancy occurs after the first hard freeze in the fall, and continues until the temperatures start warming back up in the spring. After a good hard freeze most deciduous plants lose their leaves almost immediately. It's safe to transplant trees and shrubs from that point until mid spring when they start to leaf out again. Once they leaf out in the spring they should not be transplanted.

The rules are a little different for evergreens. The transplanting window is actually longer, but to be safe I recommend that you follow the same rules for evergreens that apply to deciduous trees and shrubs. That way you won't lose any plants unnecessarily.

Overwintering Small Plants and Container-Grown Plants

Small plants, even rooted cuttings, are a lot tougher than you think!

If you have small plants in beds, rooted cuttings in flats, or plants in containers, you have to consider how to protect them from the harsh winter winds, unless of course you happen to be in a climate where you don't have harsh winter winds. In that case you still need to make sure they don't dehydrate.

First of all, you need to know what cold hardiness zone you are in. If you don't know, go to http://freeplants.com and look near the bottom of the table of contents. There you should find a link to the USDA Zone Hardiness Map.

I happen to be in northern Ohio, zone 5. If you are in zone 5, 4, or 3, everything that works for me here should work for you as well. If you are in zone 6 you'll need less protection than what I recommend here, but not that much less. If you are in zone 7 or higher, the rules change and you'll need very little, if any protection at all. Of course all of this varies depending on what it is that you are growing.

For the most part, plants in flats or containers should be covered with white plastic for the winter. **The plastic absolutely, positively, must be white. You cannot use clear plastic!** We use white plastic because we want to reflect the rays of the sun. The plastic is not to keep your plants from freezing. They are going to freeze as hard as rock, even covered with white plastic. It won't hurt them as long as they are hardy for the zone that you are growing them in. The white plastic reflects the rays from the sun, and actually keeps your plants frozen for longer periods of time once they do freeze. The damage is done during the freezing and thawing process that can occur if your plants are uncovered.

The white plastic keeps the wind off your plants and at the same time retains moisture, keeping your plants properly hydrated throughout the winter.

The reason that we don't use clear plastic is that it actually magnifies the rays of the sun, heating up the air around your plants, making them think it is spring, and they will come out of dormancy. Then at night the temperature drops down below freezing and your plants with all that soft tender growth will freeze and likely die. That's why it's important to keep your plants cool and/or frozen until spring arrives.

White plastic is hard to find, and usually the only place you can find it is at a nursery or greenhouse supply company. But you can use clear plastic and then immediately paint it white with latex paint. You just have to make sure the paint doesn't flake off during the winter, exposing areas of clear plastic that will allow the sun to peek through. You might have to go out every once in a while and touch up your paint job.

When covering your plants for the winter the plastic should not rest on your plants. You need to build a structure over your plants, and then stretch the white plastic over that. There are a number of ways to accomplish this.

One is to use concrete reinforcing mesh. Your plants should be in rows no more than 48 inches wide. Just cut pieces of mesh long enough to form a hoop over your 48-inch wide row. Overlap the pieces of mesh slightly. When you cut the mesh, cut right in the middle of the squares so you leave three-inch prongs that can be stuck in the ground to keep the mesh in place. Then just pull the plastic over the mesh and weight it down with soil on the sides and the ends.

Another method is to stand concrete blocks on end, lay small pipes on top of the concrete blocks, then lay snow fence over the pipes. Then pull the plastic over top and weight it down with soil.

Don't Forget Mouse Bait!

Field mice love all kinds of plants, especially rooted cuttings, and they will eat them right down to the ground. They also love these mini hoop houses that we build in our nurseries. Mice work hard building a nest for the winter, and when you build one of these mini hoop houses you help their cause a great deal. What more could they ask for? A nice warm environment and an endless supply of food and they are as happy as happy can be.

Don't use regular mouse bait sold in most stores, it will never hold up in the moist conditions of a mini hoop house. If you go to a farm supply store they have mouse bait that is designed for damp conditions and holds up longer.

What about the small plants in beds? I don't cover mine at all. Once I plant them in a bed I leave them to fend for themselves and they seem to do just fine. The roots stay plenty moist in the soil, and there's something magic about being planted in the ground. My small plants in beds have always gone through our Ohio winters just fine.

If you are in zone 6 or higher, you have to be careful using white plastic over the winter because it's possible that it can get too warm inside. In most cases your plants will do just fine outside with a minimal amount of protection. Just make sure to keep them watered, even during the winter. Drying out is the biggest danger, even to dormant plants. Keep in mind, even though your plants appear to be dormant, the roots are still actively growing and need a constant supply of moisture.

The Basics of Plant Propagation

From my http://freeplants.com/ site people write to me all the time and ask me, "How do I propagate this plant?" or "How do I propagate that plant?" In an effort to simplify the process so folks can achieve a greater level of success with their plant propagation efforts, this chapter will help you determine whether your plant is considered evergreen or deciduous and what plant propagation technique you should use, at what time of the year.

Keep in mind there is also the "How to Do What" chapter near the end of the book. You'll want to refer back to both of these chapters on a regular basis.

Plant propagation comes down to two things; timing and technique. Timing is critical and usually determines whether or not your efforts will be successful. The techniques are simple and fun, and if you use the correct technique at the correct time of the year, you are certain to be a successful plant propagator. Let's see if I can find a simple way of explaining it all so it makes sense.

First of all, consider the plant that you want to propagate. Is it an evergreen, or is it deciduous? A deciduous plant loses its leaves during the winter, while an evergreen does not.

When many people think of evergreens, they only think about plants like Pines, Spruce, Taxus, Junipers, and Arborvitae. They don't consider the broadleaf evergreens like Rhododendrons, Azaleas, Laurel, and many of the Euonymus varieties. Then there are plants that are sometimes considered semi-evergreen, depending on the cold hardiness zone they happen to be growing in. In zones 4 or 5 they might be deciduous, but in zones 6 or 7 they could be considered evergreen, or semi-evergreen. Some Viburnums are like that.

If you don't know for sure what zone you are in - and it's always a good idea to know that as a gardener - go to http://freeplants.com/, and in the table of contents you'll find a link to the USDA Hardiness Zone Map.

For the sake of simplicity, and if you visit the message board on my website you know I'm all about keeping things simple, any plant that retains its leaves over the winter is an evergreen, and you have to make sure you put your plants in the right category as you consider which propagation technique to use for each plant.

The next important thing for you to consider is the time of the year you are trying to propagate your plant. The techniques vary from season to season, as does the "wood" of the plant you are trying to propagate. Many plant propagation techniques use stem cuttings, and a softwood cutting hardly looks or feels like wood, but that is the term that is often used when referring to a stem cutting.

For the most part, when you are trying to propagate a plant by taking a cutting from the plant, you should be making your cutting from the new growth of the plant, or at least the growth from the current season. In other words, you should take your cuttings from the tips of the branches.

As you read this book and learn about plant propagation, you are going to see the terms "hardwood cuttings", "softwood cuttings", and semi-hardwood cuttings". You might be asking, "What in the world do those terms mean?" Let me explain using a deciduous plant as an example, something like Forsythia.

Before I do that, something just occurred to me that I might as well address right now. If you happen to be in a southern climate, as I'm sure many people who read this book will be, don't get discouraged or confused when I discuss winter months, dormant plants, frozen ground, first frost date, last frost date, or a hard freeze, not a frost. Everything discussed in this book will work for you folks in the south just as it does for us here in the north, and in most cases your plants are easier to propagate. You folks still have winter months; it's just that the air and the ground are warmer. Your plants are as close to being dormant in the winter as they will get, so if the propagation technique that I'm discussing mentions dormancy, consider the coldest weather you experience and work from there. The only thing that you need more of than we do is artificial shade, which is pretty easy to create.

Now back to explaining what the terms "hardwood cutting", "softwood cutting" and "semi-hardwood cutting" mean. During the winter months all plants go dormant. The first really hard freeze of the fall or early winter forces the plants into dormancy, and in most cases they take a little rest, or nap for the remainder of the winter. It's their resting period.

When the plants wake up in the spring, they do so with a renewed vigor and they start growing like crazy. Some make beautiful flowers, but they all start putting on new growth. This new growth is very pliable, actually soft and tender. As the growing season progresses this new growth becomes more rigid and woody. By the end of the growing season this new growth is very rigid and quite hard.

You can actually take and root cuttings of many plants at different stages of the cycle I just described. The "wood", as it is known to professional growers, changes throughout the cycle, and so must your propagation techniques.

Thus the terms softwood cuttings, semi-hardwood cuttings, and hardwood cuttings. If you are taking cuttings in the late spring you are working with softwood cuttings. The cuttings you take are very soft and pliable. I say late spring because it's really difficult to work with the new growth that is extremely soft in the early spring.

You really should wait four to six weeks for this new growth to harden off a little before you try to propagate with it. The ideal time to propagate with softwood cuttings is when you can take a cutting and not have it droop over immediately. Like I mentioned earlier, this is usually four to six weeks after the plants start growing in the spring.

Softwood cuttings are actually quite easy to root, but they are also delicate and fail easily. As the growing season goes on, the new growth becomes more and more rigid, and by midsummer the wood you use for cuttings would be considered semi-hardwood cuttings. Most evergreens can be propagated during the midsummer, during the period when the wood is considered semi-hardwood. You can do semi-hardwood cuttings starting in midsummer through late summer and early fall. Even if they are not rooted by the time winter arrives, just leave them in the propagation system until spring and they will often root as soon as the weather starts to warm up. Evergreens do not do well as softwood cuttings.

By late fall or early winter the wood becomes very hard and woody, and the cuttings you take then would be considered hardwood cuttings.

Is this making any sense to you? I hope so, because it really is important to the success of your propagation efforts. Okay, with all of that said, let's see if I can make this all come together now.

Look at the plant that you are trying to propagate. Is it an evergreen plant, or a deciduous plant? Are you trying to propagate it with softwood cuttings, semi-hardwood cuttings, or hardwood cuttings? Keep in mind, what we are discussing here for the most part applies to shrubs. Most trees are either grown from seed, or are grafted or budded onto a seedling. That goes for deciduous trees, ornamental trees and evergreens like pine and spruce.

To recap; most deciduous shrubs do well as softwood cuttings. Most evergreen shrubs do well as semi-hardwood cuttings. Some deciduous shrubs like Forsythia, Weigela, Rose of Sharon, and shrub Dogwoods can be done as hardwood cuttings. Many evergreen shrubs can also be done as hardwood cuttings, but many of them require bottom heat if you do them as hardwood cuttings.

Keep this in mind; the techniques for each category are very different. Softwood cuttings and semi-hardwood cuttings are the easiest to do, and the plants root the fastest as softwoods or semi-hardwoods. But that limits your plant propagation hobby to just a few months of the year and that's no fun! So try your hand at all of the techniques. Keep in mind that in the back of this book you'll find a chapter titled; "What You Should Be Doing Now". That chapter gives you a month by month, season by season playbook for what plant propagation techniques you should be doing at any given time of the year.

Use that chapter as your guide. Make copies of those pages (you have my permission) and hang them on your refrigerator as your reminder. In the remainder of this book you'll find a chapter for each of the plant propagation techniques that we've mentioned so far, and maybe one or two that haven't been mentioned yet.

At http://freeplants.com/ you'll find more photos of the plant propagation techniques as well as "how to" videos and a message board. I've been adding videos to my website as I've been writing this book. As of this writing I have four videos complete and there will be several more on the site before this book is published. So far people really like the videos. Check them out.

Propagation by Division

Division is a propagation technique used for landscape plants that do not have a single stem or crown. Many plants have just one stem emerging from the ground and cannot be propagated using this technique. However, there are a large number of plants that have multiple stems emerging from the ground and those types of plants can be propagated as easily as digging them up and dividing them into multiple pieces.

For the most part, perennials fall into the family of plants that have multiple stems. Examples would be hostas, mums, ornamental grasses, and many perennial flowers. Perennials die back to the ground during the winter, then re-emerge in the spring.

Division is exactly as it sounds. You just dig up the parent plant that you intend to reproduce and quite simply divide it into many plants. Of course the size of the parent plant will determine how many times you can divide it.

If you don't wish to reduce the size of the parent plant, then you would only remove a few divisions from around the edge of the parent plant. You can probably do so without actually removing the parent plant from the ground. Just scrape the ground clean around the plant so you can see exactly which sections of the parent plant you would like to remove, take a spade and force it into the ground between the parent plant and the division you are removing. Make three more cuts around the piece you are removing, completely severing the division from the parent plant, and cutting all the roots securing it in the ground. Pry the division out of the ground. Trim the roots a little if needed and replant the division in its new location.

To divide a plant into as many small plants as possible, all you do is dig up the parent plant. The entire plant will come out of the ground in one large clump. Place this clump on a hard surface and just cut it into several pieces using a large knife or even a spade. Each division should have at least two or three sprouts or eyes. This is much easier to determine in early spring once the eyes have begun to develop, but the leaves have not yet developed.

Division is a very simple form of propagation, but it is only effective with a limited number of plants. Division will not work on what would be considered the higher forms of landscape plants. If a plant has a single stem emerging from the ground, then it must be propagated by another means.

Early spring is a good time to propagate by division, however, late fall will work as well. I prefer early spring when the plants are still dormant. When you divide plants in the spring you can plant the divisions in a bed, keep them watered for a few weeks until they establish more roots, then let them grow for the remainder of the summer. You can transplant them to their permanent location in the fall if you like, but once again, I prefer to do so in the spring so they have a chance to get established before winter.

Propagation by division should be done when the parent plant is either dormant, or about to break dormancy in the spring.

The dormancy season begins in the fall after the first hard freeze. Not a frost. A frost isn't severe enough to trigger dormancy. It takes a hard freeze where the temperatures dip well below 32 degrees F. for a period of several hours. You usually know when the first hard freeze has occurred because when you wake up in the morning the leaves are falling from the trees like snow and it's obvious the leaves have been damaged. A lot of leaves start falling long before the first hard freeze, but on that day it's different. They all fall from the trees after they have been frozen! Well, maybe except the oaks. They just turn brown and hang there a while longer.

Once dormancy begins, the plants remain dormant until spring arrives and the temperatures begin to increase. As the air temperatures increase, the plants become active. Root activity can occur throughout the winter. Any time the soil temperatures rise above 45 degrees F. root growth is likely taking place below the surface of the soil. The plants are not officially out of dormancy until leaves begin to appear.

Most people don't realize that nurserymen can only dig deciduous plants while they are dormant. Once a deciduous plant develops leaves in the spring, it cannot be safely removed from the ground until late fall when it is dormant again. Digging a deciduous plant once it leafs out will immediately put the plant into shock, and likely kill it. That's why so many plants are now grown in containers. They can be sold and planted all summer long.

Evergreens are a little different. They cannot be safely removed from the ground once growth is established in the spring, but once that new growth hardens off later in the summer, it is once again safe to dig them. But to stay on the safe side I recommend transplanting when plants are completely dormant. It's less stressful on the plants.

Propagation by Layering

Layering is another simple form of plant propagation that can be done at home without special equipment. Layering is a natural form of propagation that often takes place inadvertently. Layering usually is most effective with deciduous shrubs. However, I have seen broadleaf evergreens such as Rhododendrons and Piers Japonica propagate themselves through inadvertent layering in the nursery.

Inadvertent layering is something that happens both in the natural environment of a woods, or in a commercial nursery. In the woods, the lower branches of a plant get pulled to the ground from their own weight, or because of snow sticking to or laying on the branches. As the branches lay on the ground, all of the necessary conditions exist for the plant to develop roots at the point where the branch is resting on the ground.

In a commercial nursery field, shrubs are planted in rows. The rows are cultivated on a regular basis to keep weed growth to a minimum. Through the cultivation process a few branches can be inadvertently, but partially covered with soil. When this happens the plants often develop roots at the point where they were covered with soil. Once the branch has developed enough roots to support that branch, the branch can be removed and planted as you would any other small plant. But once again, severing that branch from the parent plant can be traumatizing to the young plant, so you only want to do so when the parent plant is dormant.

At home you can intentionally layer plants to reproduce more of the same. To layer a plant you start by digging a small hole near the base of the shrub. Pull down one of the lower branches, bending it into a U, and force the bottom of the U into the hole, leaving the end of the branch sticking up out of the ground. Fill the hole with soil, covering that portion of the branch completely. The portion of the stem that is covered with soil will develop roots. Keep in mind that the end of the branch should be sticking out of the ground. You don't bury it completely, just the middle portion of the branch. See Diagram 2.

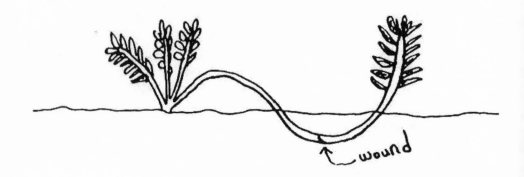

Diagram 2. The straight line in the above drawing represents the soil line. Notice how the branch is pulled below the soil line and the branch is slightly wounded at the point that is underground. The portion of the branch that is underground is covered with soil.

Making a small wound on the portion of the stem that is to be buried will help to stimulate root development. The easiest way to make a wound on the stem is to first force the branch into the hole so you know exactly what part of the branch will be resting on the bottom of the hole, then raise the stem back up and simply scrape the bark on **just one side of the stem**. All you have to do is scrape the bark enough to expose the tissue just below the bark. **Do not** wound the stem all the way around.

To further enhance your success, you can treat the wound with any kind of a rooting compound. It makes no difference whether you use liquid or powder, both work just fine. Then bury the stem as you normally would.

Some branches are more rigid than others. It may be necessary to anchor the branch down using a piece of heavy wire bent into a U, or a fork-shaped branch. Or you can hold the branch down by placing some form of weight, like a brick or a large rock on the soil covering the branch.

Layering can be done in the fall or spring. You should achieve great success with layering as late as the middle of May, but layering can really be done in any season. Layering yields great success on many deciduous shrubs because you are not actually removing the branch from the plant at the time of the layering process. The new plant that you are attempting to propagate is still attached to and being nurtured by the parent plant throughout the entire process.

Even though layering can be done until mid May, I would suggest that you do your layering as early as possible in the spring to give the plant ample time to establish an extensive root system. If you happen to do some layering later in the season you'll be fine, but it will take the plant longer to develop enough of a root system on the layered branch. So layering really knows no specific season. It's just that doing it in the off season means you will have to wait longer before you can remove the baby plant from its parent.

If you layer a plant in the fall, you should not attempt to remove that layer from the parent plant until the following fall or spring. Do not disturb the layer while the plant is actively growing. Wait until the plant is dormant.

If you layer a plant in the spring, you might be able to remove it from the parent in the fall after the growing season has ended, but I would leave it alone until spring. If you have a mild winter and the soil temperatures are fairly warm, further root development might take place over the winter months. Waiting always produces a healthier, stronger plant.

Liners planted in the field in the fall can often be forced out of the ground by the freezing and thawing process because they don't have enough time to establish enough roots to secure them in the ground. That's why I prefer to plant most of my small plants in the spring.

The term "liner" is a nursery term. Liner is short for "lining out stock". Lining out stock refers to the plants that have been grown specifically to be lined out in the field and grown on to larger sizes. A rooted cutting is smaller than a liner. A rooted cutting is just as it sounds, a four- to six-inch cutting with roots. If you plant a rooted cutting in a bed, or pot it up into a small container for one growing season, then it can be sold as a liner.

These are terms that members of our Backyard Nursery Group learn quickly. The members of our group actually supply each other with small plants to start their nurseries. Many people actually recoup all of their initial investment by selling small plants to other members of the group.

Since the invention of intermittent misting equipment, not very many commercial nurseries use layering as a means of propagation. It is too labor intensive for the number of new layers a nursery can obtain. It also ties up valuable field space that could be used for more productive things. Keep in mind that wholesale nurseries produce plants by the tens of thousands. Even members of the Backyard Growers Group often do many thousands of cuttings at a time under intermittent mist. In my Backyard Growing System I include a DVD that shows you how to build and use your own intermittent mist system.

Nurseries would have to keep a whole block of 'stock' plants that are to be used for nothing but layering, and years ago they did. They called them layering blocks. A good stock plant might yield 30 or 40 layers, where the same 'stock' plant could yield several hundred softwood cuttings. Understanding how the commercial nurserymen do layering will help you achieve better results at home.

Most deciduous shrubs have the ability to put on 18-36 inches of new growth each year. This new growth is the ideal wood for layering. If you prune the shrub heavily the year before you intend to use it for layering, it will produce a substantial amount of new growth that will be ideal for the layering process. I would actually cut the shrub to within six or seven inches from the ground. This would force the shrub to put out many new branches the following growing season. In the fall of the same season, this new growth would still be fairly pliable and could be easily pulled to the ground to be layered.

Once these new layers are rooted, they can be removed by cutting the branch just below where the new roots have developed. At the same time, the parent plant should be cut back just like before, and it will produce more new growth for the next layering season.

Serpentine Layering

You can also do what is known as Serpentine Layering. Serpentine layering is done exactly the same way as regular layering, with the exception that if the branch you are layering is long enough, you can loop it underground more than once.

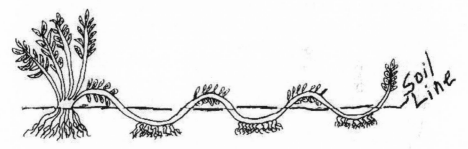

Diagram 3. Once again, the straight line represents the soil line. Notice how the branch is still attached to the parent plant, but dips below the soil line, then back up above the soil line, then once again below the soil line, and back up again. At each point where the branch dips below the soil line, the branch is wounded slighty. Roots form around the wounded area.

In order to do serpentine layering, you must leave a few buds exposed to the air and sunlight after each loop that dips underground. Each one of these loops will develop into a new plant.

Air Layering

Air layering is a propagation technique using the same theory as regular layering, except you don't bury the branch in the ground. Instead, you can use a branch up much higher on the plant, wound the stem as with regular layering, treat the wounded area with rooting compound, pack damp peat moss around that branch, and wrap it in plastic.

Air layering can be effective, but it is more time consuming. When you wrap this layered area with plastic film, you must make the seal on each end tight enough as to not allow the moisture to escape. By the same token, you must be careful not to girdle the plant by making these ties too tight.

You must keep an eye on your air layers for the duration of the time it will take to develop roots. If there do not appear to be any beads of moisture on the inside of the plastic film, you should add water to the peat moss. You should be able to do this without unwrapping the film by using a plastic cooking syringe. Just poke a hole in the plastic, add some water, and put tape over the hole when you are done.

Air layering should be done in the early spring so root development can be complete by fall. It is best to remove air layers from the parent plant before winter.

Unlike regular layers that are protected from the extreme cold by the insulation provided by snow cover, air layers would be exposed to temperatures far too extreme if left attached to the parent plant through the winter months.

There are easier ways to propagate plants than air layering, but if all else fails it's worth a try.

Softwood Cuttings of Deciduous Plants

What does the term softwood cutting mean? To help you understand the terms softwood cuttings and hardwood cuttings, let's use Burning Bush as an example.

If you watch closely as a Burning Bush develops buds in the early spring, you will see how these little tiny buds quickly develop into new growth shoots, six to eight inches in length. These new shoots develop very quickly once the plant begins to grow in the spring. In a matter of just a few weeks a little tiny bud becomes a new branch up to ten inches long. This new growth is very soft and pliable when it first emerges.

As the growing season progresses, this new growth becomes harder and more rigid. By fall this new growth has hardened off to the point that it is almost brittle, then next spring the process starts all over again. Both softwood and hardwood can be used for propagation, but the plant propagation techniques for each type of wood vary, as does the time it takes for the cuttings to root.

The only difference between a softwood cutting and a hardwood cutting is the time of year you take the cutting. Both are of the current season's growth. It is always recommended that the cuttings you use are of the current year's growth. If you go too deep into a plant to take your cuttings, you are likely to get into wood that is more than one year old. Using this older wood is almost certain to hamper your results.

Propagation of softwood cuttings is usually done at the end of spring, or early summer, depending on the climate you are in. Once plants put on new growth in the spring you have to give the new growth a few weeks to harden off to the point that the cuttings will stand on their own, and not completely wither in just a few hours after being cut. Typically you should wait from four to six weeks from the time the plants start to leaf out in the spring until you do your softwood cuttings.

Trying to do softwood cuttings prior to that is a waste of time because the wood is too soft and will wilt down very quickly. The ideal time to take softwood cuttings is just as the wood begins to harden off. I know, easy for me to say, right? - since I can tell the difference between wood that is too soft, and wood that is hard enough to work worth. Don't worry, pretty soon you too will be the expert.

Here in zone 5, northeastern Ohio, June 1st is usually our target date. Plants in this area are usually a little behind the plants in southern Ohio. We are sitting right on the southern shore of Lake Erie. When that huge body of water freezes over for the winter, it is slow to thaw and warm up. Therefore, the temperature here stays a little cooler in the spring. Of course it works just the opposite in the fall and the lake is usually responsible for sparing us from the first few frosts.

Softwood cuttings of many deciduous plants root very quickly and easily under the right conditions. However, controlling the conditions is critical. Softwood cuttings are very delicate and can dehydrate very easily, especially under the summer sun. However, with the warm temperatures of June and the tenderness of softwood cuttings, root development will occur very quickly if you can keep the cuttings from dehydrating.

The homemade plant propagation system you learned about in the beginning of this book is ideal for maintaining the humidity you need to keep softwood cuttings hydrated.

In the nursery business we use a system that actually sprays our softwood cuttings with a light mist every few minutes. In my Backyard Growing System I include a video that shows you how to build your own mist system. Details are on my http://freeplants.

com/ website. The advantage to a mist system is you can mist an entire bed of cuttings at one time.

My backyard growers root thousands and thousands of cuttings under mist each year. It gives them a huge advantage because instead of only being able to fit a hundred cuttings or so in a homemade plant propagation system, they can make and stick as many as 5,000 to 8,000 cuttings in a single day. Given that they often sell those cuttings to other nurserymen and backyard growers for as little as fifty cents each and as much as a dollar each, you can understand the advantage of having an automatic system.

But for you at home doing just a few cuttings of this or that, the homemade plant propagation system works really, really well.

Preparing a softwood cutting is easy. Just clip a cutting about four inches in length from the parent plant. Take only tip cuttings. In other words, just take one cutting from each branch, the top four inches of each branch. This is the newest growth. Strip the leaves off the lower two-thirds of the cutting, leaving just a stem and a few leaves at the top.

Wounding the cutting slightly can help the rooting process. But softwood cuttings usually root quite well without being wounded. Stripping the leaves off is wounding enough for most plants. But if you so choose, you can wound the cutting by scraping the side of the stem lightly from the bottom of the cutting up a half inch.

Visit the message board at http://freeplants.com/ and view the short video on how to do softwood cuttings.

It is always beneficial to treat your cuttings with a liquid or a powder rooting compound just prior to sticking them. Rooting compounds are available at most garden centers and do help to stimulate root development. It really doesn't matter whether you use a liquid or a powder. There are different strengths available in the powder formulas. Hardwood cuttings require a stronger formula than softwood cuttings.

Most liquid rooting compounds are sold in concentrate form and must be diluted with water. I like the liquid because all you have to do is adjust the amount of water you add, depending on whether you are doing softwood or hardwood cuttings. The instructions on the package explain how to mix the solution for softwood cuttings, semi-hardwood cuttings, and hardwood cuttings.

One of the best growing mediums for softwood cuttings is coarse sand. Typically, coarse sand is the sand that is used in the making of concrete. It is also known as sharp sand. The particles are slightly larger than, say, mason's sand or playbox sand. But it can be difficult to find since the availability of sand depends on what kind of soils you have in your area since sand is mined locally.

For that reason, many of my Backyard Growers have opted to use a mixture of peat moss and perlite or vermiculite, and they root their cuttings in flats. Just mix one part peat moss to three parts perlite or vermiculite. Another reason they prefer this peat/perlite mixture is the weight. A flat of peat/perlite mixture is much lighter than a flat of sand. Believe me, a flat of sand is quite heavy.

In the nursery we often root our cuttings in a large bed of coarse sand. But here in northern Ohio we have a good supply of coarse sand and it's easier and less expensive to fill a four foot by eighty foot bed with sand than it is a peat/perlite mixture. For one, trying to do anything with perlite on a windy day is somewhat futile.

So it really depends on what is the most practical for you. Either growing medium will work just fine. What you don't want to do is stick softwood cuttings in soil. The stems of softwood cuttings rot very easily. Most soil does not drain well enough to use as a growing medium for softwood cuttings.

The flat you use with your homemade plant propagation system should be three to four inches deep. Fill the flat to the top with your growing medium. Make your cuttings as described earlier, dip them in a rooting compound, and stick them in the flat. It helps to make a hole or a slice in the growing medium first, so the cutting will slide in easier. Softwood cuttings are not very rigid. They will break if you try to force them into the growing medium.

Using a putty knife or a masonry trowel, you can slice an opening through the growing medium, or use a large screwdriver to make a hole in the growing medium. Space your cuttings about one inch apart in the flat. Firm the growing medium around the cuttings as you stick them. You do not want air pockets around the stems. You can also water thoroughly the first time to make sure all of the voids are filled.

It is said that the ideal time to take softwood cuttings is early in the morning. However, that is not always convenient for me, so I have taken them at all hours of the day. I have never been able to determine whether or not morning, noon, or night yielded the best results. There are scientific reasons why morning is ideal, but if you don't have that option, do them whenever you can and I'm sure you'll do just fine.

Softwood cuttings wilt very quickly. Take just a few cuttings at a time and get them stuck in the growing medium and watered as quickly as possible. When you first take the cuttings, keep them in the shade for a period of seven to ten days, if possible. This gives them a chance to harden off before you put them in the sun. Plants need at least partial sun in order to develop roots.

Keep an eye on the glass or plastic cover of your homemade plant propagation system. There should be at least some water beaded up on the inside of the cover at all times. If the growing medium that you use drains really well, you'll have to water on a regular basis, but that's good because well-drained growing medium is less likely to cause problems with fungus.

If your first batch of softwood cuttings do poorly, try a new batch as soon as you realize your first batch is failing. Just a few days can make a remarkable difference in the texture of the wood as the new growth matures and hardens off a little more each day. Cuttings that wilt down almost immediately one day might do a hundred percent better two days later.

Softwood cuttings are delicate and somewhat difficult, but if you can keep them from wilting they will root very quickly. Hardwood cuttings are much more durable, but it takes considerably longer to establish roots on hardwood cuttings. Also, there are some plants that are difficult to root using the hardwood method.

An automatic mist system makes rooting softwood cuttings like child's play, and is essential for anybody who is serious about the Backyard Nursery business. As a matter of fact, when my youngest son was in the first grade, he took softwood cuttings, stripped the cuttings, dipped them in the rooting compound, and stuck them in the sand. That's all there is to it. The automatic mist system does the rest. His cuttings rooted beautifully and he later sold his plants at one of our plant sales.

Softwood Cuttings of Evergreens

The techniques and timing for making softwood cuttings of evergreens are just about the same as they are with deciduous plants, except with evergreens you have to wait a little longer for the wood to harden off before you take the cuttings. A week or two after the official start of summer is usually ideal. Then you can continue to take cuttings right up until fall. But keep in mind that as the season progresses, the new growth continues to harden off and by fall the new growth is considered hardwood. The softer the wood, the faster it will form roots.

Some of the evergreens that can be done as softwood cuttings are Taxus, Junipers, Arborvitae, Dwarf Alberta Spruce, Rhododendrons, Azaleas, Japanese Holly and Euonymus. Conifers like Blue Spruce, White Pine and Firs are easily grown from seed and don't do well as cuttings. Most trees, both evergreen and deciduous, are grown from seed and oftentimes the more desirable varieties are grafted onto those seedlings. More about that later in another chapter.

Evergreen cuttings should be stuck in coarse sand or a mixture of one part peat moss to three parts vermiculite or perlite. You can stick the cuttings in a flat and put them under the Homemade Plant Propagation System you read about in an earlier chapter. Wounding the cuttings isn't necessary because stripping the needles off the bottom third of the cutting causes enough minor injury to induce the development of callous and then roots.

Waiting until later in the fall and doing evergreens as hardwood cuttings requires less effort because they are easier to care for, but you lose the benefit of the warm temperatures available to you in July and August. As I've already mentioned, the softer the cutting wood, the faster the cuttings root. Cuttings root much more quickly if the sun can warm the growing medium to a temperature of 70 degrees F.

If you wait until the end of August you can still benefit from the warm temperatures to initiate the rooting process, and within a few weeks the temperatures will drop, reducing the amount of care required. For evergreen cuttings the ideal time is no sooner than the middle of summer through early fall. Some things are tricky and take a long time to root. The secret is to do as many cuttings as you can, and you are certain to have a reasonable degree of success.

The rules for watering softwood evergreen cuttings are about the same as for deciduous plants. Evergreens don't require quite as much water, but they still need care on a regular basis. If you put them in the Homemade Plant Propagation Unit described in this book, just make sure there are at least some beads of water on the inside of the propagator. If you use a growing medium that drains well, you can water without being concerned about them being too wet.

As with deciduous plants, intermittent mist works really great for softwood evergreens as well. But intermittent mist really isn't practical unless you want to do thousands at a time. Almost all of my backyard nursery growers use intermittent mist and many of them quickly pay for their misting systems by selling the cuttings they root to other members of our Backyard Growers Group. My Backyard Growing System comes with a DVD that shows you how to build your own Intermittent Mist System.

Hardwood Cuttings of Deciduous Plants

There are two different ways to do hardwood cuttings of deciduous plants. Is one better than the other? I really don't know. It depends on what plants you are attempting to root, what the soil conditions are at your house, and what Mother Nature has up her sleeve for the coming winter. I have experienced both success and failure using both methods. Only experimentation will determine what works best for you. Try some cuttings using both methods.

When doing hardwood cuttings of deciduous plants, you should wait until the parent plants are completely dormant. As mentioned earlier, this does not happen until you have had a good hard freeze where the temperature dips down below 32 degrees F. for a period of several hours. Here in northeastern Ohio this usually occurs around mid November.

If you are in a warmer climate where you normally don't experience freezing temperatures, wait until your plants are as close to dormant as possible. Also keep in mind that using the method described below that suggests burying your cuttings is probably not the method you want to use in warmer climates. See Method Two below.

Method One

When we discussed softwood cuttings of deciduous plants, I told you to take tip cuttings from the ends of the branches only. That rule does not apply to hardwood cuttings of deciduous plants. For instance, a plant such as Forsythia can grow as much as four feet in one season. In that case, you can use all of the current year's growth to make hardwood cuttings. You might be able to get six or eight cuttings from one branch.

Grapes are extremely vigorous. A grapevine can grow up to ten feet or more in one season. That entire vine can be used for hardwood cuttings. Of course with grapevines there is considerable space between the buds, so the cuttings have to be much longer than most other deciduous plants. The average length of a hardwood grapevine cutting is about twelve inches and still only has three or four buds.

The bud spacing on most other deciduous plants is much closer, so the cuttings only need to be about six inches in length.

Since hardwood cuttings must be done during the winter months, you probably will want to work in your garage or basement where it is not quite so cold. Of course there are still some nice days after the first freeze when working outside is possible.

Making a deciduous hardwood cutting is quite easy. Just collect some branches (known as canes) from the parent plants. Clip these canes into cuttings about six inches long. Of course these canes will not have any leaves on them because the plant is dormant. Most people would look at them and call them dead sticks, but that's hardly the case. If you examine the canes closely you will see little bumps along the cane. These bumps are bud unions, or nodes, as they are often called. It is from these bud unions that next year's growth will appear. From each of these bud unions, one or more buds will emerge the following spring and grow into a branch. That's why it's important to note where the bud unions are, and not damage them as you make your cuttings.

Diagram 4. The above drawing represents a hardwood cutting of a deciduous plant. The little black spots are the bud unions. Notice how the cut at the bottom, or butt end of the cutting is cut close to the bud union, but not into the bud union? And the cut at the top of the cutting is cut about ¾-inch above the bud union, and at an angle?

When making a hardwood cutting of a deciduous plant, it is best to make the cut at the bottom, or the butt end of the cutting just below a node, and make the cut at the top of the cutting about ¾-inch above a node. Making your cuts at those locations on the cutting serves two purposes. One, it makes it easier for you to distinguish the top of the cutting from the bottom of the cutting as you handle them. It also aids the cutting in two different ways.

Any time you cut a plant above a node, the section of stem left above that node will die back to the top node. So if you were to leave a half inch of stem below the bottom node, it would just die back anyway. Having that section of dead wood underground is not a good idea. It will interfere with callous build up, which is what happens before the roots start to form, and it also provides a place for insects and disease to hide.

So it's critical that you make the cut on the bottom of the cutting just below the bottom node, but at the same time being careful not to damage the bottom node. You need that bottom node intact and healthy in order for roots to form.

It is also helpful to actually injure a plant slightly when trying to force it to develop roots. When a plant is injured, it develops a callous over the wound as protection. This callous buildup is necessary before roots will develop. Some growers actually scrape one side of the cutting, near the bottom. They just scrape the bark off the cutting on one side, causing a wound on the side of the cutting near the bottom. Do not make this wound more than ¾-inch long. See Diagram 5.

Diagram 5. This drawing actually represents a softwood cutting because it has leaves. But notice how the cutting has been wounded on one side near the bottom?

Making the cut on the top of the cutting at ¾-inch above the node, as you saw in Diagram 4, is done so that the ¾-inch section of stem above the node will provide protection for the top node. This keeps the buds from being damaged or knocked off during handling and planting. You can press down on the cutting without harming the buds. Although not necessary, it helps to make the cut at the top of the cutting at an angle for two reasons. One, when the top cut is made at an angle it sheds water away from the cut end of the cutting and helps to keep disease and insects away from the cuttings. Secondly, it helps you easily distinguish the top of the cutting as you handle your cuttings. It's important to know the top from the bottom because they won't grow planted upside down.

Nurseries often hire teenagers to help with the planting of cuttings, and I can assure you, thousands of cuttings have been planted upside down by teenagers who were just mindlessly going through the motions. I once had my son and my nephew potting bare root dormant Japanese Maples, only to come back and find two plants potted upside down! The tops were buried and the roots were sticking out of the pot.

I thought I'd include that little tidbit in this book because they are both likely to be raising teenagers in years to come, and they need not forget how teenagers think.

Once you have all of your cuttings made, dip them in a rooting compound. Make sure you have the right strength rooting compound for hardwood cuttings. As mentioned earlier, powder rooting compounds come in different strengths, and you need the stronger formula for hardwood cuttings. Softwood cuttings require a weaker formula. It will tell you right on the package what kind of cuttings that particular formula is for. Most liquid rooting compounds come in a concentrated formula and then you add water and dilute it to the strength that you need, depending on what you are rooting and at what time of the year.

Don't get all caught up in brand name or liquid versus powder. They all basically do the same thing, and it's important to keep in mind that the rooting compound is not a magic potion when it comes to rooting cuttings. In many cases cuttings will root just fine without any rooting compound at all. All the rooting compound does is enhance your chances of success. The magic is to make sure you are using the correct technique at the proper time of the year for the particular plant that you are doing. This book will help you achieve that, and of course the message board at http://freeplants.com/ is a great place to come for any questions that you might have.

You can find rooting compounds at most full service garden centers, or just go to your favorite internet search engine and type in "rooting compound" and order online.

Once you have your cuttings made, line them up so the butt ends are even and tie them into bundles. Select a spot in your garden that is in full sun. Dig a hole about ten inches deep and large enough to hold all of the bundles of cuttings. (If you are doing grapes, the hole will have to be deeper because the cuttings are so long.) Place the bundles of cuttings in the hole **upside down.** The butt ends of the cuttings should be pointing up, near the top of the hole. Put the cuttings in the hole so the butt ends are about six inches below the surface of the ground before you backfill the hole.

Carefully put some soil around the cuttings, backfilling around the sides of the cuttings right up to the point where the soil is even with the butt ends of the cuttings. Don't cover the butt ends just yet. Press down the soil around the sides firmly so it's not likely to settle much at all. At this point all you should be able to see are the butt ends of the cuttings.

Mix some peat moss with water in a pail until the peat is very soggy. You might even want to mix the peat moss and water several hours before you need it because peat can be difficult to wet thoroughly, then just let it sit for several hours, making sure the peat is thoroughly wet. Pack about two inches of the soggy peat over the butt ends of your cuttings, then finish backfilling the hole with soil. Mark the location with a stake, so you can find your cuttings in the spring when it's time to plant them in your garden.

Over the winter the cuttings will develop callous and possibly some roots. Placing them in the hole upside down puts the butt ends closest to the surface, so they can be warmed by the sun, creating favorable conditions for root development. Being upside down also discourages top growth.

Don't disturb the buried cuttings until about mid spring, after the danger of frost has passed. Over the winter the buds will begin to develop and will be quite tender when you dig them up. Frost would do considerable damage if you dig them and plant them out too early. That's why it is best to leave them buried until the danger of frost has passed. Here in zone 5 northern Ohio we can get a frost as late as May 15, so that's usually my target date to dig up my hardwood cuttings and get them planted right-side up.

Dig them up very carefully, so as not to damage them. Cut open the bundles and examine the butt ends. Hopefully, you will see some callous buildup. Even if there is no callous, plant them out anyway.

You don't need a bed of sand or anything special when you plant the cuttings out. You can just put them in a sunny location in your garden. Of course the area you choose should be well drained, with good rich topsoil.

But to really enhance your chances of success with your hardwood cuttings, put them in the Homemade Plant Propagation System you learned how to build at the beginning of this book.

If you are doing thousands of hardwood cuttings, stick them in a bed of sand under your intermittent mist system. My **Backyard Growing System** comes with a DVD that shows you how to build your own intermittent mist system. Intermittent mist is like magic when it comes to rooting cuttings and you can use it from early spring right up until winter, doing tens of thousands of cuttings if you like. You can find more details about intermittent mist on my website, http://freeplants.com/.

Once you dig up your hardwood cuttings in the spring, planting them is very easy. Just make a slice in the soil so you can easily push them into the soil without breaking off any roots that may have already formed. Plant the cuttings two to three inches deep, making sure that you have at least two or three buds above ground. Backfill around the cuttings with loose soil, then water thoroughly to make sure there are no air pockets around the bottom of the cuttings. That's all you need to do. Then water them lightly every few days if you have them planted out in your garden. If you have them in the **Homemade Plant Propagation System**, just make sure there are beads of water on the inside of the glass indicating that you have a high level of humidity.

Within a few weeks the cuttings will start to leaf out. Some will more than likely collapse because there are not enough roots to support the plant. The others will develop roots as they leaf out. By fall, the cuttings that survived should be pretty well rooted. You can transplant them once they are dormant, or you can wait until spring. If you wait until spring, make sure you transplant them before they break dormancy, keeping in mind that timing is the most important factor when you are propagating plants.

Many deciduous plants grow easily from hardwood cuttings, some do not. Hardwood cuttings are not likely to work for some of the more refined varieties of deciduous ornamentals. Check the **"How to Do What"** chapter of this book to see if the plant you want to do is listed. If it's not there, just pop into the message board at http://freeplants.com/ and ask.

There are many wholesale nurseries where the employees spend a great deal of their time during the winter months making hardwood cuttings using this method. Years ago I worked for a nursery that produced hundreds of thousands of grape plants each year. The owner of the nursery expected us to produce 5,000 cuttings each day. Of course he would give us about two weeks to work our way up to that figure. Once we were able to make 5,000 a day, he would then put us on piece work. We earned eight hours' wages for 5,000 cuttings. Eventually I was able to make 5,000 cuttings in six hours, but it was a very fast pace.

Method Two for Hardwood Cuttings of Deciduous Plants

When using the second method for rooting hardwood cuttings of deciduous plants you do everything exactly the same as you do with Method One, up to the point where you bury them **upside down** for the winter.

With Method Two you don't bury them at all. Instead, you plant the cuttings out as soon as you make them in the late fall. In other words, you just completely skip the step where you bury the cuttings underground for the winter.

Plant them exactly the same way as described for Method One. As with all cuttings, treating them with a rooting compound prior to planting will help induce root growth.

You can do hardwood cuttings of deciduous plants throughout most of the winter, as long as the ground is not frozen.

As I mentioned at the beginning of this chapter, whether or not you choose to bury your hardwood cuttings of deciduous plants is up to you. It might be fun to experiment and bury some and not others. Keep in mind that softwood cuttings of all plants are usually easier to do, but at least hardwood cuttings give you something to do during the winter months.

Don't forget to check the **"What You Should Be Doing Now"** chapter to make sure you don't miss any propagation techniques that you could be doing now.

Hardwood Cuttings of Evergreens

Hardwood cuttings of evergreens are usually done after you have experienced two heavy frosts in the late fall. Here in zone 5 Ohio that happens around mid November or so. But I have obtained good results with some plants doing them as early as mid September, taking advantage of the warmth of the fall sun. Try some cuttings early and if they do poorly, just do some more in November.

If you are doing cuttings of evergreens before you experience a hard freeze, the plants are not dormant and are considered semi-hardwood cuttings. Semi-hardwood cuttings will root faster than hardwood cuttings, but they need more care, such as daily watering if you have them outside in a bed of sand. But if you are using the Homemade Plant Propagation System, you won't have to water anywhere near as often.

Hardwood cuttings of many evergreens can be done at home in a simple frame filled with coarse sand. And to make this as simple as possible, almost any sand will work, but the coarser the sand the better. By communicating with my Backyard Nursery Growers I've learned that many people have a difficult time finding coarse sand. It's out there, and in most areas of the country it's the same sand used to make concrete. So you have to go to a gravel yard, gravel pit, or concrete supply company to find it.

If you still can't find it, you can use bagged play box sand, or swimming pool filter sand. I've even had people use sand-blasting sand. You can also use a mixture of peat moss and perlite. Use one part peat moss to at least three, maybe four parts perlite or vermiculite. You can find perlite and vermiculite at most full service garden centers.

Make a frame using 2- by 6-inch boards to contain your growing medium. Nail the four corners together as if to make a large picture frame. This frame should sit on top of the ground in an area that is well drained. An area of partial shade is preferred.

Once you have the frame constructed, remove any weeds or grass inside the frame so vegetation does not grow up through your propagation bed. Fill the frame with coarse sand or the growing medium of your choice. The growing medium in the frame must be well drained. Standing water is sure to seriously hamper your results.

Making the evergreen cuttings is easy. Just clip a cutting four to five inches in length from the parent plant. Take tip cuttings only. Strip the needles or leaves from the bottom one-half to two-thirds of the cutting. Wounding evergreen cuttings isn't usually necessary because removing the leaves or needles causes enough injury for callous buildup and root development.

Dip the butt ends of the cuttings in a powder or liquid rooting compound and stick them in the sand about ¾- to one inch apart. Keep them watered throughout the fall until cool temperatures set in. Start watering again in the spring and throughout the summer. They don't need a lot of water, but be careful not to let them dry out. And at the same time make sure they are not soaking wet.

Hardwood cuttings of many evergreens will root this way, but it does take some time. You should leave them in the frame for a period of twelve months. You can leave them longer if you like. Leaving them until the following spring would be just fine. They should develop more roots over the winter.

A friend of mine who is a wholesale nurseryman uses this method to root almost all of his evergreens. He covers his frames with steel hoops and plastic to provide some extra protection over the winter. This can help, but you must be careful. Do not use clear plastic. Clear plastic allows too much sun to penetrate and it will get too hot on the nice days and the plants will start to come out of dormancy too early. Then they will freeze when the temperature dips back down below freezing. If you are going to cover your frame for the winter, use white plastic or clear plastic that has been whitewashed with white latex paint mixed with water. You must also water during the winter if you are going to cover the plants with plastic. Dehydration occurs very easily during the winter.

For the home gardener I recommend not covering the frame for the winter. A covering of light fluffy snow actually protects plants from harsh winter winds. Let Mother Nature take care of your cuttings over the winter. Sometimes she does a fantastic job, and sometimes she reminds us that we are tinkering with nature.

This method of rooting hardwood cuttings can and will work for a variety of different evergreen plants, both needled and broadleaf evergreens. But there are some varieties that are more difficult and will not root unless special care is provided. For most of the more difficult to root evergreens, the addition of bottom heat will help to induce root development. We will discuss bottom heat in another chapter.

Keep in mind that any time we attempt to root a cutting of any kind, we are asking the plant to establish roots before the top of the plant starts growing. Once the plant begins to grow it will die if it has not established roots first. Softwood cuttings are very delicate and will collapse if not cared for carefully. However, softwood cuttings root very quickly and can be growing on their own roots in a matter of a few weeks. Hardwood cuttings, on the other hand, are much more durable and can survive for months with very little care or roots. However, hardwood cuttings are very slow to develop roots.

Yes, it can be confusing, but it's also fun to master each of the techniques as you progress and learn.

Semi-Hardwood Cuttings of Evergreens

Most cuttings of evergreens don't do well as softwood cuttings, but if you wait until the middle of summer you can do many evergreens as semi-hardwood cuttings. Make the cuttings exactly as you would if you were doing hardwood cuttings of evergreens. Make the cuttings four to five inches long, strip the needles or leaves from the bottom two-thirds of the cutting, dip them in a rooting hormone and stick them in a flat or bed in the growing medium of your choice.

With semi-hardwood cuttings of evergreens you either have to put them in the **Homemade Plant Propagation System,** or put them under intermittent mist. They will never hold up in a bed out in the open like hardwood cuttings do.

The advantage of doing them as semi-hardwood cuttings is they root much more quickly. If you do them in mid to late summer they are rooted before winter.

Using Bottom Heat to Root Cuttings

Landscape plants love heat. When they are warm they grow. When they are cool they do not grow. Of course as with everything else, extreme heat is not good.

We know that root growth can and will begin taking place once soil temperatures reach or exceed 45 degrees Fahrenheit. Even if the top of a plant is not active, the roots can be actively growing underground if the soil is adequately warm. Professional growers have learned that if you can heat the soil or other growing medium without raising the air temperature around the tops of the plants, root development can be induced and/or speeded up.

This technique is used by professional propagators around the world. The same evergreens that require up to twelve months to develop roots, as discussed in previous chapters, can be successfully rooted in as little as six weeks by applying bottom heat. The secret to using bottom heat is to warm the soil without increasing the air temperature above the soil. In a commercial nursery the ideal situation is to maintain a soil temperature of 69-70 degrees F. and an air temperature of 40-45 degrees F.

Nurserymen use different ways to accomplish this. One of the most popular ways is to set the propagation frames on benches about 36 inches high inside a greenhouse. Using a small forced-air furnace, they blow the warm air under the benches. The benches have plastic or some other material around the bottom so the heat cannot escape out the sides. The heat must rise up through the growing medium. A special thermostat is inserted in the soil. This thermostat controls the furnace. When the soil reaches the optimum temperature, the heat is turned off until it is needed again.

Trapping the heat under the bench keeps the air temperature much lower than the soil temperature. Therefore, rooting activity can take place while the top of the cuttings remain dormant.

Professional growers also use very complex systems that circulate hot water through plastic lines buried in the soil. This works well, but a second heating system is usually required to keep the air around the top of the cuttings from getting too cold.

Nurserymen are innovative people. Some of them have used regular household water heaters, installed a circulating pump, and circulated hot water through one-inch plastic piping. When you consider a hot water boiler sells for $1,500 or more, this is a great idea. A new household water heater can be purchased for less than $200. Nurserymen also often find used counterflow furnaces that have been removed from homes, and they use them in their greenhouses to force hot air under benches to create bottom heat. A counterflow furnace is a furnace that pushes the air out the bottom of the furnace instead of the top.

For you and me at home, these systems are too complex. After all, we're only interested in rooting a few cuttings or maybe a few hundred, not fifty or sixty thousand. So how can we use bottom heat to increase our results and to cut down on the amount of time it takes to root our cuttings?

An easy way to create bottom heat in your propagating frame would be to purchase an electric soil-warming-cable kit, or a heating mat. I like the heating mats because all you have to do is plug them in and set the flat of cuttings onto the mat. Soil-warming cables work, but it's a lot more trouble to bury them in the flat or frame that you are using. Soil heating mats come in a variety of different sizes. You might have trouble trying to find one locally, but the gardening mail-order catalogs have them, or you can just use the internet to find one.

Make sure your heating mat comes with a built-in thermostat that regulates the temperature of the soil automatically. These thermostats are preset and cannot be changed. The ideal soil temperature for rooting most cuttings is 69 degrees Fahrenheit.

Read the instructions and make sure you are using the mat correctly.

Bottom heat can also be used to induce and speed the rooting of hardwood cuttings of deciduous plants. Purple Sandcherry is an extremely popular landscape plant. Purple Sandcherries can be grown from softwood cuttings, but even then they can be tricky and fail easily.

But they can also be grown from hardwood cuttings. One of our local growers discovered that if he tied all of his hardwood cuttings in bundles, then placed the entire bundle in sand or another growing medium and applied bottom heat to the butt ends of the hardwood cuttings for about fourteen days before sticking the cuttings, the cuttings would actually callous up quickly and a much larger percentage would eventually root.

Make your hardwood cuttings of deciduous plants as described earlier in this book, tie them in bundles, and place them right-side up in a bed of coarse sand or other growing medium equipped with bottom heat for a period of 14-20 days. The cuttings will develop callous and be ready to plant out with a much higher degree of success.

Some plants are extremely difficult to root using other methods. Rhododendrons, for instance, are very slow to root, if they root at all using other methods, but with bottom heat they root quite fast. The same thing holds true for Taxus, some Junipers, and some Arborvitae.

Intermittent Mist

Intermittent mist is used primarily for softwood and semi-softwood cuttings of both deciduous plants and evergreens.

As you learned earlier in the section regarding softwood cuttings, they will root very quickly under the right conditions, but softwood cuttings are extremely delicate and will expire quickly if not properly tended to. The Homemade Plant Propagation System does a great job of keeping softwood and semi-hardwood cuttings hydrated, but if you want to do more than, say, 150 cuttings at a time, it's not very practical.

Intermittent mist is like hiring a full-time nanny for your softwood cuttings. An intermittent mist system automatically applies a very small amount of water every few minutes, all day long, until your cuttings are rooted. Even though intermittent mist is off and on throughout the day, the actual amount of water used is quite low. The water nozzles are very small, allowing only a minimal amount of water to pass through. When the water is on, the duration of spray is very short, using very little water. For a typical day misting as many as 10,000 cuttings, the actual "on" time would be less than seventeen minutes for the entire day. Given that the misting nozzles are very tiny, the actual amount of water used is minimal.

Once the sun goes down the mist system is turned off, giving the cuttings a chance to dry out overnight. Using intermittent mist also reduces the chance of fungi that grow in more humid environments. Even during the day when the mist is on, the cuttings still have an ample amount of air circulation. That's the one disadvantage of the Homemade Plant Propagation System, not enough air circulation. But it still works really well if you don't have intermittent mist.

My **Backyard Growing System** comes with a DVD that shows you how to build an intermittent mist system identical to the one that I use. The nice thing about the intermittent mist system that I show you how to build is that it's fully automatic. All you do is turn it on, stick your cuttings and walk away. It comes on in the morning as scheduled, keeps your cuttings misted all day, and then shuts itself off in the evening just before the sun goes down, giving your cuttings a chance to dry out before bedtime.

Cuttings root like magic under intermittent mist. Really. Some in as little as two weeks, four weeks for most others.

With intermittent mist you can root as many as 5,000 cuttings at a time in an area the size of a picnic table. We stick as many as 8,000 cuttings in a day to put under the mist. Many of my backyard growers quickly pay for their mist systems selling only rooted cuttings.

When you first take your softwood cuttings, it is a good idea to mist them more frequently, then after a few days you can change the settings and mist them less often. The mist system allows you to control how often you want to mist the cuttings, and for how long. I explain all of that in the video, and give you the times that I typically use.

When I first start sticking my softwood cuttings around the first week of June, which is about six weeks after the plants break dormancy in the spring, the cuttings are very soft and delicate, so delicate that I only make a small amount of cuttings and immediately get them under the mist. By a small amount I mean a few hundred at a time.

When lunchtime comes around, if we have a few hundred cuttings that we've taken but aren't quite ready to stick, I just toss them in a flat and place the entire flat under the mist while we eat lunch. The mist system acts like a nanny, keeping them fresh until we get back to work.

Can you tell that I get excited about intermittent mist?

All you have to worry about is a power failure. If for some reason the electricity supply to your home is temporarily interrupted, you should keep the cuttings watered by hand until the power is restored. Of course this is only a problem during the daylight hours. If it happens to be an overcast day and there is no direct sunlight on the cuttings, they won't need much water. On the other hand, if it is a hot sunny day, watering is extremely important.

When you take cuttings and prepare them for propagation under intermittent mist, follow the exact same procedures described in the previous section on softwood cuttings.

Another advantage of intermittent mist propagation is if you experience failure with certain cuttings, you can try another batch two or three weeks later. You don't have to wait months to know whether or not you were successful.

The new growth on a landscape plant matures so quickly that a matter of just a few days can make all the difference in the world. One day the cuttings might be too soft to survive, but a few days later the wood will have hardened off to the point that they will do just fine. If you stick some cuttings and they are too soft and wilt down immediately, just a few days later the new growth of the very same plant may have hardened off enough to yield cuttings that will do quite well.

Propagation is often hit or miss. Everything we know about plant propagation has been learned through trial and error. The problem is that the people who know all the answers are tight lipped. The professional propagators in the nursery business spend eight hours a day, fifty two weeks a year making baby landscape plants, millions and millions of plants. They constantly try new ideas. Some of them work and some don't, but just like any other professional in any other industry, they don't give away their trade secrets. Sometimes it's really funny. I know dozens of people in the nursery business in this area. Many of them are friends. Some are more than willing to share their knowledge, but you won't believe how secretive some of them can be when I ask them how to do something. They just clam up and look at me like I'm from Mars.

I can assure you that writing this book and making it available to homeowners is not going to set well with some people in the nursery business. I know they are not going to appreciate me teaching people about mist propagation.

Intermittent mist was developed primarily to aid in the propagation of softwood cuttings, but I use my mist system from late spring right up until it starts to freeze. As the season progresses and the cuttings require less water, I adjust the system accordingly. As the days get shorter I adjust the system to come on later and go off earlier. I just keep reducing the hours of operation until the system is only on three to four hours each day.

I made some Rheingold Arborvitae cuttings in mid September and by the time I was closing my nursery up for the winter the cuttings were completely rooted. This was at a time of the year when it is usually too late for softwood cuttings and too early for hardwood cuttings, but by taking advantage of the heat of the sun and using intermittent mist to keep the cuttings watered lightly, I was able to achieve terrific success.

Intermittent mist is incredible, but as soon as you start using it you are going to have a lot of questions. And without a place to get expert advice quickly, you will experience failures that can really set you back. I don't want that to happen. Thus the creation of the private group for Backyard Growers. This is an internet-based community where you can communicate with me on a daily basis as well as hundreds of others who are also in the Backyard Nursery business. You can learn more about that at http://freeplants.com/.

Growing Plants from Seed

Many landscape plants can be grown from seed, but with many different plants it is much quicker and easier to grow plants from cuttings than it is to grow them from seed. Seedlings are so tiny and delicate when they first germinate that they require much more care than a cutting.

Many plants will not come true from a seed. In other words, the seedlings produced from seeds that were collected from a red Rhododendron are not likely to flower red. More than likely the flowers will be a pale lavender. Plants grown from seeds collected from a pink Dogwood will most likely flower white.

However, there are certain plants that must be grown from seed. Most Taxus varieties are successfully grown from cuttings, but Taxus Capitata, one of the most attractive varieties of Taxus, will not come true from a cutting. However, it will come true from seed.

Taxus Capitata is the most popular variety of the pyramidal-shaped Taxus. Cuttings from this plant can be rooted and will grow just fine, but the plant will not have the natural pyramidal shape of the parent plant. The plants of this variety grown from cuttings tend to grow more upright and require much more pruning to obtain the desired pyramidal effect.

Other plants that are routinely grown from seed are plants that are extremely difficult to grow from cuttings. Many ornamental trees either cannot be grown from cuttings, or if they are grown from cuttings, the plants have weak root systems. Most trees are grown from seed. Not all will be true to the parent plant, but in many cases the only way to get a true clone of the parent plant is by budding or grafting a piece of the desired variety onto a small tree grown from seed.

In other words, ornamental trees will not come true from seed, but the plants that are produced from seed have good strong root systems and the same basic genetic makeup of the desired variety. Therefore those seedlings can be used as rootstock to produce the desired variety. The desired variety is grafted onto the healthy, hardy rootstock of the seedling. Grafting will be discussed in detail in the next section of this book.

Growing landscape plants from seed is a little more difficult than growing vegetables. The seeds produced by most landscape plants will not germinate until they have undergone certain environmental conditions. Most seeds from landscape plants have a very hard, protective outer shell. Under natural conditions most of these seeds do not receive the proper treatment in order for the seeds to germinate. They just lay on the ground and either dry out or rot.

In different climates, different varieties of plants will grow naturally from seed. In eastern Pennsylvania for instance, Rhododendrons and Mountain Laurel grow wild on the mountain sides. Here in northern Ohio, dogwoods grow wild. Of course only a fraction of the seeds produced in the woods actually make it to the point of germination and survival.

Many seeds go through a period of internal dormancy right after the fruit falls from the tree. In some cases, there is actually a chemical barrier that prevents the seeds from germinating while they are still inside the fruit.

The hard protective coating on some seeds was designed by nature to protect the seed, but in many cases this protective coating actually inhibits the germination of the seed because water and air cannot penetrate the hard coating. Or at least it takes a long time for this hard outer coating to soften to the point that it can absorb enough moisture and oxygen to initiate germination. And often by the time the seed is ready to germinate, the timing is all wrong, and even though germination takes place, the weather conditions at the time might be unfavorable for the survival of the small seedling.

Many seeds actually require a double dormancy period before they will germinate. In other words, the seeds must lay on the ground completely dormant for one full growing season, and then germinate the following growing season. During the first season the only thing that is taking place is the outer coating is being softened by the elements. Once the outer coating is softened, water and air can penetrate and germination can begin.

Timing is critical. Once the protective coating is softened and the seeds begin to receive sufficient amounts of oxygen and water to begin germination, the plant will start to grow. However, if this takes place at the wrong time of the year, the young seedling will be destroyed by the intense summer sun or the freezing temperatures of winter. That's why, of the millions of seeds produced by landscape plants, so few actually germinate and survive to become adult plants.

As a gardener you can control when your seeds will germinate by putting them through a pretreatment process commonly known as stratification. You can actually fool some seeds into germinating much more quickly by creating the necessary environmental conditions to soften the outer coating and initiate germination sooner. This usually involves both moistening the seeds, and putting them through a cold treatment.

As an example let's consider the process of "tricking" a Japanese Red Maple seed into germinating on "our terms", where we control the timing.

Japanese Red Maple trees start producing their seeds during the summer. By late fall the seeds are fully mature and ready to be harvested. Harvesting the seeds too soon will ruin your chances of successfully getting them to germinate because the embryo is just not completely formed. You should wait as late into the fall as possible before picking the seeds. Keep an eye on them. During the summer they will be green in color, by fall they will start to turn a reddish brown. At that point they are probably ready to be picked. The best way to know when they are ready to harvest is when they start falling from the tree on their own.

I should also mention that not all Japanese Maples are created equal. If you want Japanese Maple seedlings that will have really deep red color, you should collect your seeds from a Japanese Red Maple tree that holds its red color late into the summer and fall. That still doesn't mean that all of your seedlings will have that same deep red color, because as you learned earlier, seedlings do not always come true to the parent plant. But it does increase your odds of getting some really nice red seedlings.

So where do you find Japanese Red Maple trees that you can collect seeds from? You have to look around and pay close attention in your daily travels. If you live in an area that is considered zone 4 through zone 7 by the United States Department of Agriculture, there should be an abundant supply of Japanese Red Maple trees in your area. You might find a really nice tree in a park, a cemetery, in the landscape of a public building, or in the front yard of a private home. Ask for permission to pick a few seeds and be on your way!

Put the seeds in a paper bag and store them in a cool dry place until you are ready to start the pretreatment process. Make sure you put them in a paper bag and not plastic. A plastic bag can trap moisture and create a more humid condition than you want while you are storing the seeds.

Next you have to decide what day you want to plant your seeds or tiny seedlings outside. You certainly don't want tiny seedlings popping up in the early spring when you are still likely to experience a late freeze that would kill them. Here in zone 5, northern Ohio we often experience late freezing as late as early May. So to be on the safe side I use a target planting date of May 15 or May 20. Where you live the date is likely to be different, but the date for planting seedlings outside is the same as the date you would consider it safe to plant delicate annual flowers outside without being concerned about them freezing.

So as an example we'll use May 15. Next count backwards on the calendar 90 days from that date. So with our example, that would take us back to February 15. The pretreatment process for our Japanese Maple seeds is going to take 90 days, so if we start the process on February 15 our seeds will be ready to be planted outside on May 15, which puts us past the date of the last freeze in our area.

When you harvest your seeds in the fall, keep them stored in a paper bag in a cool dry place until you are ready to start the pretreatment process. At that time you will remove the seeds from the bag and break off the little wing. Japanese Maple seeds look just like regular maple tree seeds, except they are smaller. The actual seed is all the way to one end of the seed pod, so the rest of the seed pod - the wing - can be snapped off and discarded.

Put the seeds into a plastic freezer bag with about four times as much peat moss as you have seeds. You don't have to use peat moss; you can also use perlite, vermiculite, or sand. Whatever medium you decide to use should be moist, but not soaking wet. Mix the seeds with the medium, poke a few holes in the bag so some air can circulate, then place the bag in your refrigerator for a period of 90 days. Don't put the bag way to the back of your refrigerator, it might be too cold back there and the seeds and medium could freeze. Even though this won't hurt the seeds, it will slow down the necessary process.

After about 60 days you can start checking the seeds about once a week, looking for signs of germination. You'll see a little sprout coming out of the seeds when they start to germinate. Once about ten percent of your seeds have started to germinate, you can then sow the entire batch of seeds in a flat of peat moss and perlite mixture. Use three parts perlite to one part peat moss. You can also use a seed-starting potting mix instead of the peat/perlite mixture, if you prefer.

When you sow the seeds in the flat, just barely cover them with soil. The rule of thumb for the planting depth of seeds is twice the length of the seed, which isn't very deep at all. After planting, water thoroughly, then allow the growing medium to dry out before watering again. If you keep the growing medium too wet, it will remain cool and slow the germination process and increase the chance that the seeds might rot before they germinate. You could also consider shining a lamp on the flat to warm the soil as you wait for the seeds to germinate. Or you can place the flat on a heating mat made for germinating plants.

Watering is critical. You don't want the seeds and the medium to get too dry, but at the same time too wet is also a problem.

When you have the seeds in the refrigerator, another option is to use tweezers to just pick out the seeds that have started to germinate and plant only the seeds that have shown signs of germination. This gives the slower seeds more time in the cold treatment process. As with anything that has to do with plants, there are all kinds of options.

Another technique for the pretreatment process is to just place the seeds on a damp paper towel, making sure the entire paper towel is damp and pressed against the seeds. Fold the towel over the top of the seeds, and place the paper towel within a plastic bag in the refrigerator.

Another option is to actually scarify the seeds before you start the pretreatment process by nicking the outer coating of the seed with a knife, file, or small saw blade. Nicking the outer coating of the seed actually creates a place where the moisture can penetrate faster, speeding the germination process. Some professional growers actually put the seeds through a process where they expose the seeds to an acid treatment to scarify them. I do not recommend this at all. I'm sure it works if you know what you are doing, but it is dangerous and not something you want to do at home.

Growing trees like Japanese Red Maples and Flowering Dogwoods from seed is fun and exciting. And like many other forms of plant propagation, it can also be addicting. But that's a good thing since it is also very therapeutic.

If you would like to grow dogwoods from seed, the process is pretty much the same, except you need to remove the seeds from the pulp as soon as you harvest them. Dogwood trees produce a seed that is inside of a fruit, and you must remove the seed from the fruit before you place them in the paper bag for storage. If you soak the seeds, fruit and all, in a bucket of water for a few days, this will soften the fruit and you can squeeze the fruit between your fingers, forcing the seeds out of the fruit. You can do this while the seeds and fruit are still in the water, and once you have them separated just add water to the bucket slowly until it gently overflows. The seeds should lay on the bottom of the bucket while what is left of the fruit and pulp should float up and out of the bucket.

Both White Dogwoods (cornus florida) and Chinese Dogwoods (cornus Kousa) can be grown from seed. The fruits are very different, but the process is the same.

On my website, http://freeplants.com/ I have a couple of short "how to" videos, one on growing dogwoods from seeds, and one on growing Japanese Maples from seed. What you just read will make more sense if you watch the videos.

Many trees and shrubs can be grown from seed, and most need some form of a pretreatment process, but it does vary from plant to plant. Check the "How to Do What" section of this book for details about the plant you'd like to grow. If you don't find what you need there, go to the internet and do a search for "seed germination database". I'm sure you'll find the answer there. But you'll need the botanical name of the plant you are searching for in order to get the data you need.

Grafting

Grafting is one of the most interesting forms of plant propagation. It is also one of the most tedious and least used forms of plant propagation. Many wholesale nurserymen stay away from grafting because it is just too labor intensive. They either will not grow plants that have to be grafted, or they will buy small grafted plants from someone who specializes in grafting. Don't let that scare you off. Nurserymen are in business to make money. If it takes too long to produce a particular plant, they either don't grow it at all, or they buy small plants from another grower. Or in many cases they buy plants that are landscape size and ready to sell and just re-wholesale them.

If you are daring enough to try your hand at grafting, you will realize a tremendous amount of pride and self satisfaction when you successfully graft your first plant. Grafting is not difficult; it just takes patience and the correct conditions.

One of the most beautiful landscape plants on this planet is the Laceleaf Weeping Japanese Red Maple (acer palmatum dissectum). This tree is very low growing; most are not more than four feet tall. The branches spread out, making the tree wider than it is tall. The branches weep from the top of the tree to the ground, the foliage is deep red in color and the leaves are delicately cut on the edges. The plant is breathtaking during the spring and summer months. Nobody walks by this plant without taking notice. It is just as interesting during the winter. The weeping branches create a very unique effect even though the plant is without leaves.

The only truly effective method of propagation for the Laceleaf Weeping Japanese Red Maple is grafting. Very few nurseries grow them. That's why a three-foot-tall plant in a garden center is likely to have a price tag of $150 or more, often commanding a price of $300 or more. You can grow one yourself for a little bit of nothing.

Grafting is the art of attaching a piece of one plant to another in such a way that the two pieces will bond together and become one plant. One plant is used to provide the root system and sometimes the stem, and is commonly known as the rootstock. A piece of the desired plant is grafted onto the rootstock. The small piece of the desired plant is known as a scion, which is really nothing more than a cutting, as we learned earlier in the book.

In the case of the Laceleaf Weeping Japanese Red Maple, the rootstock would be a Japanese Maple grown from seed. The Laceleaf Weeping Japanese Red Maple would be the desired variety. A cutting (a scion) is taken from the Laceleaf Japanese Maple and grafted onto the rootstock. That scion will bond with the rootstock and continue to grow into a tree that will have all of the identical characteristics of the original tree that you wanted to clone.

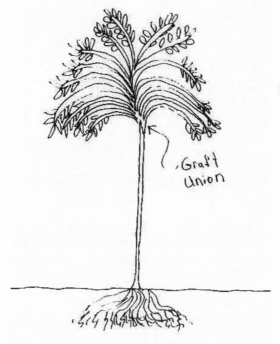

Diagram 6. In the above drawing notice the point of

> **the graft union? On this particular plant the graft union is up high. Sometimes the graft union is at ground level. Everything below the graft union is rootstock, and everything above the graft union is the desired variety.**

Any new growth that happens to come from below the graft union would be considered suckers and should be removed. Those suckers will have all of the characteristics of the rootstock and not the scion, therefore making them undesirable.

Before you can create a Laceleaf Weeping Japanese Maple through the magic of grafting, you must first raise a regular Japanese Maple tree from seed to use as the rootstock. You can do this using the techniques described in the section on growing landscape plants from seed. The ideal size of the rootstock for grafting Japanese Maples is 3/16- to ¼-inch in diameter, which is usually a two- or three-year-old seedling.

Grafting is typically done in January or February. In the late fall pot up the seedlings that you intend to use as rootstock for grafting that winter. Use a good quality, well-drained potting soil. Many bagged soils do not drain as well as they should so you should consider mixing in some additional perlite or vermiculite. Keep these potted plants outside, but in a protected area until about January 15. Make sure they do not dry out. Plants need moisture during the winter as well as during the growing season. You must leave them outside, so they remain dormant up until the time you are ready to graft them.

You can build a wooden frame outside and cover it with white plastic to help protect your Japanese Maple rootstocks. Put weatherproof mouse bait in with the plants. Field mice can and will chew on the stems of your plants, girdling them and causing them to die.

White plastic reflects the sun. Don't use clear plastic, it will get too warm inside when the sun is out, the plants will start to break dormancy, and then sustain damage when the temperature dips below freezing at night. When storing plants for the winter you want them to stay at one constant temperature. Or at least as constant a temperature as possible.

Once you bring them inside, you should let them warm up for a period of two to three weeks before you start grafting. Keep them at a temperature of 70 degrees F. After about fourteen days the plants should start showing signs that they are beginning to break dormancy. At this point they should be grafted immediately. The piece of the desired variety that is to be grafted onto the rootstock should remain outdoors in the cold (completely dormant) right up until the day you are going to graft. You don't want this part of the plant trying to grow until the graft is at least partially healed.

In order to achieve success with grafting you need to understand exactly what part of the plants you must bond together. There is a thin layer of tissue sandwiched between the bark of the tree and the wood. This tissue is known as the cambium layer. You can liken the cambium layer of a tree to the circulatory system in your own body. The cambium layer transfers water and nutrients to the top of the plant from the roots and vice versa.

When grafting it is extremely important that you bond the cambium layer of the rootstock with the cambium layer of the scion. (The scion is the term used to describe the piece of the desired plant variety that you are attaching to the rootstock.) Matching up these two surfaces as closely as possible is extremely important. The two sections of cambium layer are going to bond and will be the only thing holding the plant together. This bond is almost like a natural form of welding if you want to look at it from a mechanical point of view.

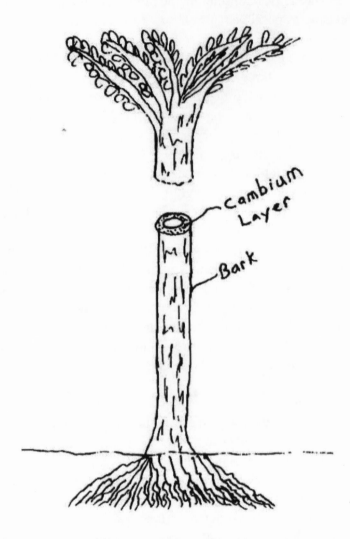

Diagram 7. This drawing shows a cross section of a typical tree. Notice how the cambium layer is between the bark of the tree and the wood that makes up the center of the tree.

In the above drawing you see where the cambium layer is located in relation to the bark of the tree and the center wood of the tree. It is important to understand that the wood of a tree is no longer a viable part of the tree and if you insert the scion below the cambium layer and against the wood of the tree, the graft will not bond. You must match the cambium layer of the scion to the cambium layer of the rootstock.

Diagram 8. This type of graft is called a Saddle Graft.

There are many different kinds of grafts, but all are based on the same basic theory. Match up two compatible plants and bond the two cambium layers together. Over the years, through trial and error, nurserymen have learned what plants are compatible with other plants are far as what plants can be used as rootstock. Yet I still have not found a resource that clearly explains what is compatible with what. But here is some of what I've learned over the years:

Japanese Maples must be grafted to Japanese Maples.
Lilac can be budded or grafted to a privet rootstock.
Cotoneasters can be budded or grafted to a Washington or Paul's Scarlet Hawthorne rootstock.

Most other plants should be budded or grafted onto plants that are genetically the same just so you are not wasting your time. In other words, use a White Pine rootstock to graft a Weeping White Pine onto.

Performing the actual task of making the graft union is not that difficult. The secret is to make sure that as you cut into the cambium layer, you do not cut too deeply and into the wood. Make sure the scion wood and the rootstock are as close to the same size diameter as possible. If they are different sizes, the cambium layers will not line up and the grafts will not be successful. If you are doing a veneer graft, the rootstock and the scion don't have to be the same size because it is still possible to match up the cambium layers. There are grafting photos on my website, and by the time this book is published I should have a short grafting video on the site as well. http://www.freeplants.com/

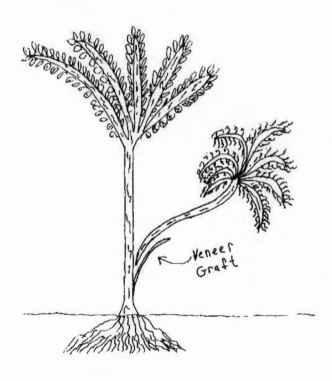

Diagram 9. This type of graft is called a Veneer Graft. The scion is inserted just below the bark, between the bark and the cambium layer.

Once you have joined the plants together, the graft union should be wrapped with a rubber band to firmly hold the twigs in place. After the rubber band is in place, the entire graft union should be coated with melted grafting wax to keep the union airtight. If air gets into the graft union, the cambium layers will dry out and not bond. Make sure the grafting wax is not too hot. Just warm enough for it to melt is as hot as you should let it get. If the wax is too hot, tissue damage can occur. The rubber band should be left on for a period of about eight weeks.

Caring for recently grafted plants after the process is complete is extremely important to the success of your efforts. Once the graft is complete, keep the plants warm; 70 degrees F. is ideal. Maintain this temperature for a period of at least three to four weeks, giving the graft unions plenty of time to heal.

Maintaining a relatively high humidity around the graft union also helps the healing process. One way to do this is to wrap the graft union with a piece of plastic and make sure some moisture gets trapped under the plastic.

Make sure your plants also receive some light. Natural light from a window is best, but if that is not possible, provide some artificial light.

Don't move your new grafts outside until the danger of frost has past. Be careful not to put them in the full sun right away. At least 50% shade is best until they harden off completely. I keep my grafted plants under shade for the entire first season. I also provide winter protection for them during their first winter outside. You can create shade by building a wooden frame and laying snow fence on top of the frame.

Grafting is not difficult to do, but it does require patience and an area where you can work indoors during the winter. Grafting is well worth the effort because of the incredible satisfaction of seeing your very first grafted plant thrive. Of course your friends and family will probably think you're a little nutty creating "Frankenstein" plants in your little laboratory. At least they may think that until they see what you've created!

Budding

Budding is another form of grafting, except with budding you do not attach a small branch of the desired variety. You only insert a single bud under the bark of the rootstock.

Budding is a mid to late summer project, usually done around the end of July or the beginning of August. It is at this time of the year that the bark of the young trees will slip. In other words, the bark is somewhat loose from the tree and a bud can be slipped between the bark and the cambium layer easily.

Budding is easier than grafting and is used quite often in the nursery industry. Almost all flowering crabapples are propagated through the budding process.

The rootstock is grown from seed using the techniques described earlier. Once the rootstock reaches ¼-inch in diameter, the budding is done. A small 'T'-shaped cut is made in the bark of the rootstock, the bark is gently pulled away from the cambium layer with a knife, but only enough to allow a single bud to be slipped under the bark.

The bud you are going to insert under the bark of the rootstock will be removed from a branch of the variety you would like to grow. You can remove a small branch from the desired variety. This branch is called a bud stick. This bud stick can have as many as twenty or more usable buds on it. Each bud has a leaf attached to it. Pinch the leaf off but leave the leaf stem. This stem will serve as a handle as you work with the bud. If you have a hundred different crabapple rootstocks that you grew from seed, you can grow many different varieties of flowering crabapples by inserting different varieties of buds into these rootstocks.

After you have made the "T" cut in the rootstock and loosened the bark slightly, you are ready to remove the desired bud from the bud stick. The bud is removed by slicing into the branch under the bud you are removing. Almost like peeling an apple, except you cut below the bark and remove a piece of bark along with the cambium layer attached. The bud is still attached to the piece of bark and cambium you remove. Do this carefully and do not cut into the bud and damage it. By the same token, you don't want to cut too deeply into the wood. If you cut into the wood, your bud will not take when inserted into the rootstock. Just like grafting, you have to have cambium tissue against cambium tissue in order for the bud to successfully bond to the rootstock.

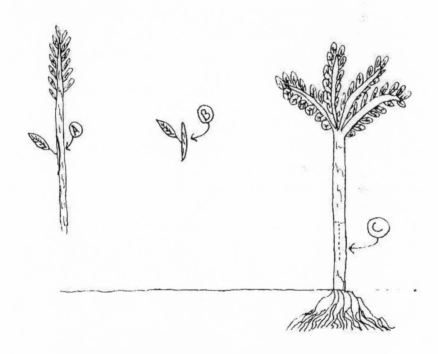

Diagram 10. "A" represents the bud stick from which the buds are removed. "B" the bud and leaf after being removed from the bud stick. "C" indicates where the bud will be inserted under the bark of the rootstock.

As soon as the bud is removed from the parent plant, it should be immediately inserted under the bark of the rootstock. Once inserted, the bud union should be wrapped securely with a rubber band so the bonding process can begin. The rubber band should be tight enough to close up the flaps that are holding the bud in place, but not so tight that it girdles the tree. Nothing further should be done this growing season. Just let Mother Nature take over until spring.

Early in the spring while the plant is still dormant, the rootstock should be cut off just above the inserted bud. When the plant breaks dormancy the bud will begin to grow into a plant identical to its parent plant.

Budding is a much simpler form of grafting because you can do it during the summer months and do not have to provide artificial heat or protection for the plant over the winter months. Budding does not work for all plants, but it is used on a wide variety of fruit trees, crabapples, some maples, and many other ornamentals.

The bud is usually inserted into the rootstock right at ground level, but in some cases, where the desire is to create a weeping tree, the bud is often inserted up high, usually about six feet off the ground. In this case it's common to insert more than one bud to make the tree develop the weeping head faster. One bud is inserted on each side of the stem. But of course the root stock is much larger and this is easier to do.

How to Do What

Please note: In this chapter I often suggest using intermittent mist. If you don't have an intermittent mist system, just use The Homemade Plant Propagation System described in the beginning of this book at the same time of the year.

Arborvitae: Cuttings taken in mid to late summer can be rooted in coarse sand under intermittent mist. Cuttings taken in the fall can be rooted in coarse sand in an outdoor frame. Cuttings taken during the winter can be rooted in coarse sand with bottom heat.

Ash Trees: Collect the seeds when they ripen in the fall and plant them out right away. Most should germinate the first season.

Azaleas, deciduous varieties: Most deciduous Azaleas are grown from seeds collected in the fall and planted immediately. I would sow them in a flat, in an area where they can be kept warm and receive some natural or artificial light. You can also try softwood cuttings, preferably under intermittent mist.

Azaleas, evergreen: Most growers do evergreen Azaleas in the late fall with bottom heat. You can try softwood cuttings in the late spring or early summer.

Barberry: Most varieties of Barberry can be done by either softwood cuttings in early June, or hardwood cuttings in the late fall.

Boston Ivy: Grow from seed. Plant outdoors in late April or early May.

European Beech: Grow from seed. Collect when ripe in the fall, plant outdoors immediately.

Purple Leaf Weeping Beech: This variety must be grafted onto a Beech variety grown from seed.

White Birch: Grow from seed. Collect the seeds when ripe and plant outdoors in the fall.

Weeping White Birch: This variety must be grafted onto a Birch rootstock grown from seed.

Boxwood: Softwood cuttings in July under intermittent mist or hardwood cuttings in mid to late fall in an outdoor frame. Winter cuttings with bottom heat.

Burning Bush: Softwood cuttings in late May or early June, hardwood cuttings in late fall to winter in an outdoor frame.

Weeping Cherries: Weeping cherries must be grafted onto a cherry rootstock grown from seed. Collect the seeds when ripe, stratify 150 days over winter, plant in the spring. I have also had some success with softwood cuttings under intermittent mist.

Blue False Cypress: Semi-hardwood cuttings in late August under intermittent mist, or hardwood cuttings in the late fall with bottom heat.

Gold Thread Cypress: Hardwood cuttings in late fall with bottom heat. You can try some semi-hardwood cuttings in late summer under intermittent mist.

Clematis: Softwood cuttings in late spring. As with almost all softwood cuttings, intermittent mist will dramatically increase your success.

Cotoneaster: Softwood cuttings in early June, or hardwood cuttings in late fall.

Flowering Crabapples: Most varieties of flowering crabapple must be grafted or budded onto a rootstock grown from seed. Collect the seeds as they ripen in the fall and plant them outdoors immediately.

Daylilies: Propagate by division in the fall or the spring. Some growers do them during the summer.

Chinese Dogwood: Softwood cuttings in early June or grow from seed. Collect the seed in the fall when ripe. Stratify in moist peat in your refrigerator for 90 days before sowing.

Pink Dogwood: Softwood cuttings under intermittent mist in early June, or bud or graft onto a white dogwood seedling.

Red Twig Dogwood: Layering in April or May, or softwood cuttings in June, or hardwood cuttings in late fall.

Yellow Twig Dogwood: Layering in April or May, or softwood cuttings in June, or hardwood cuttings in late fall.

Variegated Dogwood Trees: Softwood cuttings under intermittent mist in early June, or bud or graft onto a white dogwood seedling.

White Dogwood: Softwood cuttings in early June or grow from seed. Collect the seed in the fall when ripe. Stratify in moist peat in your refrigerator then plant outside.

English Ivy: Softwood cuttings during the summer beginning in early June.

Variegated Euonymus Varieties: Softwood cuttings beginning in June. Hardwood cuttings in the fall outside in a frame of course sand.

Firethorn (Pyracantha):   Softwood cuttings in June, or semi-hardwood cuttings in the fall.

Fir, Concolor: Grow from seed. Collect the seeds in the fall and store them in a cool dry place until spring. Sow the seeds outdoors in the spring. Cover the seed bed with clear plastic until the seeds begin to germinate.

Forsythia: Layering in spring or fall, softwood cuttings in June, hardwood cuttings in the late fall or winter.

Washington Hawthorn: Grow from seed. Collect the seeds in the fall and plant them in an outdoor seed bed immediately.

Canadian Hemlock: Grow from seed. Collect the pine cones in the fall before they open and release the seeds into the air.   Place the pine cones in a paper bag to catch the seeds as the cones open. Store the seeds in a cool dry place until spring, stratify for 30 days in moist peat in your refrigerator, and plant outside after the danger of frost has passed.

English Holly: Semi-hardwood cuttings in mid summer. Hardwood cuttings, late fall with bottom heat.

Japanese Holly: Medium softwood cuttings in mid summer, or hardwood cuttings in the fall in an outside frame of sand. Or hardwood cuttings in late fall or winter with bottom heat.

Honeysuckle: Layering in the spring, softwood cuttings in early June, or hardwood cuttings in the fall.

Hosta: Propagate by dividing in late fall or early spring.

Blue Hydrangea: Softwood cuttings, or division.

P.G. Hydrangea: Layering in the spring, or softwood cuttings in early summer.

Junipers: Softwood to semi-hardwood cuttings in mid to late summer under intermittent mist, hardwood cuttings in the fall in an outdoor frame, or hardwood cuttings in late fall or winter with bottom heat.

Leucothoe: Softwood cuttings in June or hardwood cuttings in the fall.

French Lilacs: Lilacs must be budded or grafted onto a rootstock grown from seed. Either a lilac seedling as a rootstock, or some growers use privet.

Miss Kim Lilac or Common Lilac: Softwood cuttings in late spring.

Linden Trees: Grow from seed. Collect the seeds when ripe and plant immediately.

Lirope: Propagate by division.

Magnolia: Some varieties are grown from seed, and others are budded onto these seedlings.

Maple Trees: Grow from seed. Collect the seeds when ripe and plant immediately.

Japanese Maple: Grow from seed. Collect the seeds when ripe and store until late fall. Pre-treat the seeds by soaking overnight in hot water, and then stratify in moist peat for 90 days in your refrigerator. Then plant them outside.

Weeping Japanese Maple: This variety must be grafted onto a rootstock grown from seed.

Mockorange: Layering in the spring, softwood cuttings in June, and hardwood cuttings in the fall and winter.

Mountain Ash Trees: Grow from seed. Collect when ripe and plant immediately.

Blue Myrtle: Propagate by division or softwood cuttings in late spring/early summer.

Oak Trees: Grow from seed. Collect when ripe and plant immediately.

Ornamental Grasses: Propagate by division.

Pachysandra: Propagate by division, or softwood cuttings.

Bradford Pear Trees: Grow from seed. Collect when ripe and stratify in moist peat in your refrigerator for 60-90 days.

Flowering Plum Trees: Desired varieties must be budded onto a rootstock grown from seed. Collect the seeds when ripe and stratify in moist peat in your refrigerator for 150 days before planting outside.

White Pine Trees: Grow from seed. Collect the pine cones in the fall before they open and allow them to open in a paper bag to catch the seeds. Store in a cool dry place until spring, then sow them outside.

Weeping White Pine: Must be grafted onto a white pine seedling.

Austrian Pine: Grow from seed. Collect the pine cones in the fall before they open and allow them to open in a paper bag to catch the seeds. Store in a cool dry place until spring, then sow them outside.

Mugho Pine: Grow from seed. Plant them outside in the spring.

Potentilla: Softwood cuttings in June, or hardwood cuttings in the late fall.

Poplar Trees: Grow from seed. Collect the seeds when ripe and plant outside immediately. Also softwood cuttings or hardwood cuttings.

Purple Leaf Winter Creeper: Softwood cuttings in early June, or semi-hardwood cuttings throughout the summer.

Pussy Willow: Layering in the spring, softwood cuttings in early June, or hardwood cuttings in the late fall.

Privet: Layering in the spring, softwood cuttings in early June, or hardwood cuttings in the late fall.

Redbud Trees: Grow from seed. Collect when ripe and plant outside in the spring.

Rhododendrons: Can be grown from seed. Collect in the fall and grow in a flat indoors at 70 degrees F. with some light. Hybrid varieties must be grown from cuttings. Softwoods in early June under intermittent mist, or hardwoods in a perlite/peat moss mixture in the late fall with bottom heat.

Rose of Sharon: Layering in the spring, softwood cuttings in early June, or hardwood cuttings in the late fall.

Purple Sandcherry: Layering in the spring, softwood cuttings in early June, or hardwood cuttings in the late fall.

Spirea: Layering in the spring, softwood cuttings in early June, or hardwood cuttings in the late fall.

Dwarf Alberta Spruce: Softwood cuttings in mid to late June under intermittent mist, or hardwood cuttings in the late fall with bottom heat.

Colorado Blue Spruce: Grow from seed. Collect the pine cones in the fall before they open and allow them to open in a paper bag to catch the seeds. Store in a cool dry place until spring, then sow them outside.

Viburnum: Layering in the spring, softwood cuttings in early June, or hardwood cuttings in the late fall.

Weigela: Layering in the spring, softwood cuttings in early June, or hardwood cuttings in the late fall.

Wisteria: Layering in the spring, softwood cuttings in early June, or hardwood cuttings in the late fall.

Weeping Willow: Layering in the spring, softwood cuttings in early June, or hardwood cuttings in the late fall.

Witch Hazel: Layering in the spring, softwood cuttings in early June, or hardwood cuttings in the late fall.

Yews (Taxus): Semi-hardwood cuttings in mid summer or hardwood cuttings in the fall in coarse sand in an outside frame, or hardwood cuttings in late fall or winter with bottom heat.

Yucca: Propagate by taking cuttings from the roots in early spring and planting outside. Just cut a piece of root about ¾-inch long and plant it below the surface of the soil about a half inch.

What You Should Be Doing Now

January: (mid winter)

You can do hardwood cuttings of deciduous plants. Just wait for a day when the ground is not frozen so you can either plant them out, or bury them as described in the section on hardwood cuttings.

You can also do hardwood cuttings of evergreens, if you can provide them with some bottom heat.

If you are going to do any grafting, now is the time to bring in the rootstock and let it warm up and start coming out of dormancy.

February: (mid to late winter)

You can still do hardwood cuttings as described for January. Start your grafting toward the middle or end of the month.

March: (late winter, early spring)

It's a little late for hardwood cuttings of evergreens, but you can still do some hardwood cuttings of deciduous plants. As soon as the ground thaws and spring begins to peek around the corner you can start doing plants that can be propagated by division. You can also start to do some layering.

If you have landscape plants that need pruning, do it now before they begin to grow. Any transplanting that you intend to do should be done now before the plants break dormancy.

April: (early spring)

There are plenty of things to do in April. You can still do some division as long as the plants are not too far out of dormancy. You can do layering, air layering and serpentine layering. If you have seeds that you have been stratifying, you can plant them out as long as they have been in stratification for the proper length of time.

April is also the time to start thinking about an intermittent mist system. Don't wait until the last minute to order the equipment. You want to be all ready to go when the cuttings are ready to be taken. Details can be found at http://freeplants.com/.

May: (mid spring)

You can continue all methods of layering. All seeds should now be ready to plant out. By the end of the month you should be able to start some softwood cuttings, unless you are in a northern state.

June: (early summer)

By now you should be able to do softwood cuttings of just about all deciduous plants. If you are going to do softwood cuttings of Rhododendrons, try some early in June. If they don't do well, try a few more later in the month. If you are using intermittent mist you can experiment with all kinds of different plants. June is a little early to be doing softwood cuttings of evergreens but you can test a few.

July: (mid summer)

Continue with softwood cuttings of deciduous plants. Now is the time to start some softwood cuttings of evergreens. By mid to late July you can start budding dogwoods, apples, crab apples, cherries, and anything else you would like to bud.

August: (mid to late summer)

Continue with softwood cuttings of evergreens. By now the wood of most deciduous plants has hardened off. You can still make cuttings with this harder wood if you are using intermittent mist, but you should use a little stronger concentration of rooting compound. Budding can be done early in August. Especially dogwoods, they can be budded later than most other plants.

September: (late summer, early fall)

Start watching for fall seeds to ripen and start collecting them. Evergreen cuttings can still be taken and rooted under intermittent mist. If you are not using mist, you can stick them in a bed of sand and keep them watered, or put them in the Homemade Plant Propagation System.

October: (fall)

Hardwood cuttings of evergreens can be stuck in a bed of sand. Or you can start sticking hardwood cuttings of evergreens using bottom heat. After a good hard frost you can start dividing perennials. Collect pines cones from Pines, Spruce, and Firs. As the cones open they release the seeds inside. Store the seeds in a cool dry place until spring for planting. Seed pods from Rhododendrons and Deciduous Azaleas can also be collected.

November: (late fall)

Hardwood cuttings of evergreens can be stuck either in a bed of sand outdoors, or indoors with bottom heat. Hardwood cuttings of deciduous plants can be done by either of the methods mentioned in the section on hardwood cuttings.

If you intend to do some grafting over the winter, now is the time to make sure your rootstock is potted up and placed in a protected, but cold area until January.

December: (early winter)

You can do hardwood cuttings of evergreens in a bed of sand or with bottom heat. You can also do hardwood cuttings of deciduous plants as long as the ground is not frozen.